Steel- and Toolmaking Strategies and Techniques before 1870

Hand Tools in History Series

- Volume 6: Steel- and Toolmaking Strategies and Techniques before 1870
- Volume 7: Art of the Edge Tool: The Ferrous Metallurgy of New England Shipsmiths and Toolmakers from the Construction of Maine's First Ship, the Pinnace *Virginia* (1607), to 1882
- Volume 8: The Classic Period of American Toolmaking, 1827-1930
- Volume 9: An Archaeology of Tools: A Catalog of the Tool Collection of the Davistown Museum
- Volume 10: Registry of Maine Toolmakers
- Volume 11: Handbook for Ironmongers: A Glossary of Ferrous Metallurgy Terms: A Voyage through the Labyrinth of Steel- and Toolmaking Strategies and Techniques 2000 BC to 1950

Steel- and Toolmaking Strategies and Techniques before 1870

H. G. Brack

Davistown Museum
Publication Series Volume 6
© Davistown Museum 2008
ISBN 978-0-9769153-4-8

ISBN 13: 978-0-9769153-4-8
ISBN 9: 0-9769153-4-0
Davistown Museum

Front Cover illustration:
 "Fig. 3 Oinoche, British Museum, reg. no. 1846.6-29.45" from Oddy, W. A., and Judith Swaddling. 1985. Illustrations of metalworking furnaces on Greek vases. In: Craddock, P. T., and M. J. Hughes, eds. *Furnaces and smelting technology in antiquity*. British Museum Occasional Paper 48, London. pg. 54.

Cover design by Sett Balise

Pennywheel Press
P.O. Box 144
Hulls Cove, ME 04644

Acknowledgements

This publication was made possible by a donation from Barker Steel LLC.

Many thanks to The British Museum, Museum of London, Maison de l'Outil, Ironbridge Gorge Museum Trust, The Metals Society, Amsterdam Archaeology Depot, John D. Rockefeller Jr. Library, New Bedford Whaling Museum, Dave Brown, and Thorndike Library.

Also present and endlessly helpful through all stages of the *Hand Tools in History* series and deserving of much credit for it are Judith Bradshaw Brown, Linda Dartt, and Beth Sundberg, without whose typing, editing, filing, and patience the books would not have come to be. And thanks also to Sett Balise, whose technical skills have greatly enhanced my research and the availability of the museum's oeuvre in cyberspace.

Preface

Davistown Museum *Hand Tools in History*

One of the primary missions of the Davistown Museum is the recovery, preservation, interpretation, and display of the hand tools of the maritime culture of Maine and New England (1607-1900). The *Hand Tools in History* series, sponsored by the museum's Center for the Study of Early Tools, plays a vital role in achieving the museum mission by documenting and interpreting the history, science, and art of toolmaking. The Davistown Museum combines the *Hand Tools in History* publication series, its exhibition of hand tools, and bibliographic, library, and website resources to construct an historical overview of steelmaking techniques and strategies and the edge toolmakers of New England's wooden age. Included in this overview are the roots of these strategies and techniques in the early Iron Age, their relationship with modern steelmaking technologies, and their culmination in the florescence of American hand tool manufacturing in the last half of the 19th century.

Background

During 38 years of searching for New England's old woodworking tools for his Jonesport Wood Company stores, curator and series author H. G. Skip Brack collected many different tool forms with numerous variations in metallurgical composition, many signed by their makers. The recurrent discovery of forge welded tools made in the 18th and 19th centuries provided the impetus for founding the museum and then researching and writing the *Hand Tools in History* publications. In studying the tools in the museum collection, Brack found that, in many cases, the tools seemed to contradict the popularly held belief that all shipwrights' tools and other edge tools used before the Civil War originated from Sheffield and other English tool-producing centers. In many cases, the tools that he recovered from New England tool chests and collections dating from before 1860 appeared to be American-made rather than imported from English tool-producing centers. Brack's observations and the questions that arose from them led him to research the topic and then to share his findings in the *Hand Tools in History* series.

Hand Tools in History Publications

- Volume 6: *Steel- and Toolmaking Strategies and Techniques before 1870* explores ancient and early modern steel- and toolmaking strategies and techniques, including those of early Iron Age, Roman, medieval, and Renaissance metallurgists and toolmakers. Also reviewed are the technological innovations of the Industrial Revolution, the contributions of the English industrial revolutionaries to the evolution of the factory system of mass production with interchangeable parts, and the development of bulk steelmaking processes and alloy steel technologies in the latter half of the 19th century. Many of these technologies play a role in the florescence of American ironmongers and toolmakers in the 18th and 19th century. Author H. G. Skip Brack cites archaeometallurgists such as Barraclough, Tylecote, Tweedle, Smith, Wertime, Wayman, and many others as useful guides for a journey through the pyrotechnics of ancient and modern metallurgy. Volume 6 includes an extensive bibliography of resources pertaining to steel- and toolmaking techniques from the early Bronze Age to the beginning of bulk-processed steel production after 1870.

- Volume 7: *Art of the Edge Tool: The Ferrous Metallurgy of New England Shipsmiths and Toolmakers* explores the evolution of tool- and steelmaking techniques by New England's shipsmiths and edge toolmakers from 1607-1882. This volume uses the construction of Maine's first ship, the pinnace *Virginia*, at Fort St. George on the Kennebec River in Maine (1607-1608), as the iconic beginning of a critically important component of colonial and early American history. While there were hundreds of small shallops and pinnaces built in North and South America by French, English, Spanish, and other explorers before 1607, the construction of the *Virginia* symbolizes the very beginning of New England's three centuries of wooden shipbuilding. This volume explores the links between the construction of the *Virginia* and the later flowering of the colonial iron industry; the relationship of 17th, 18th, and 19th century edge toolmaking techniques to the steelmaking strategies of the Renaissance; and the roots of America's indigenous iron industry in the bog iron deposits of southeastern Massachusetts and the many forges and furnaces that were built there in the early colonial period. It explores and explains this milieu, which forms the context for the productivity of New England's many shipsmiths and edge toolmakers, including the final flowering of shipbuilding in Maine in the 19th century. Also included is a bibliography of sources cited in the text.

- Volume 8: *The Classic Period of American Toolmaking 1827-1930* considers the wide variety of toolmaking industries that arose after the colonial period and its robust tradition of edge toolmaking. It discusses the origins of the florescence of American toolmaking not only in English and continental traditions, which produced gorgeous hand tools in the 18th and 19th centuries, but also in the poorly documented and often unacknowledged work of New England shipsmiths, blacksmiths, and toolmakers. This volume explicates the success of the innovative American factory system, illustrated by an ever-expanding repertoire of iron- and steelmaking strategies and the widening variety of tools produced by this factory system. It traces the vigorous growth of an American hand toolmaking industry that was based on a rapidly expanding economy, the rich natural resources of North America, and continuous westward expansion until the late 19th century. It also includes a company by company synopsis of America's most important hand toolmakers working before 1900, an extensive bibliography of sources that deal with the Industrial Revolution in America, special topic bibliographies on a variety of trades, and a timeline of the most important developments in this toolmaking florescence.

- Volume 9: *An Archaeology of Tools* contains the ever-expanding list of tools in the Davistown Museum collection, which includes important tools from many sources. The tools in the museum exhibition and school loan program that are listed in Volume 9 serve as a primary resource for information about the diversity of tool- and steelmaking strategies and techniques and the locations of manufacturers of the tools used by American artisans from the colonial period until the late 19th century.

- Volume 10: *Registry of Maine Toolmakers* fulfills an important part of the mission of the Center for the Study of Early Tools, i.e. the documentation of the Maine toolmakers and planemakers working in Maine. It includes an introductory essay on the history and social context of toolmaking in Maine; an annotated list of Maine toolmakers; a bibliography of sources of information on Maine toolmakers; and appendices on shipbuilding in Maine, the metallurgy of edge tools in the museum collection, woodworking tools of the 17th and 18th centuries, and a listing of important New England and Canadian edge toolmakers working outside of Maine. This registry is available on the Davistown Museum website and can be accessed by those wishing to research the history of Maine tools in their possession. The author greatly appreciates receiving information about as yet undocumented Maine toolmakers working before 1900.

- Volume 11: *Handbook for Ironmongers: A Glossary of Ferrous Metallurgy Terms* provides definitions pertinent to the survey of the history of ferrous metallurgy in the preceding five volumes of the *Hand Tools in History* series. The glossary defines terminology relevant to the origins and history of ferrous metallurgy, ranging from ancient metallurgical techniques to the later developments in iron and steel production in America. It also contains definitions of modern steelmaking techniques and recent research on topics such as powdered metallurgy, high resolution electron microscopy, and superplasticity. It also defines terms pertaining to the growth and uncontrolled emissions of a pyrotechnic society that manufactured the hand tools that built the machines that now produce biomass-derived consumer products and their toxic chemical byproducts. It is followed by relevant appendices, a bibliography listing sources used to compile this glossary, and a general bibliography on metallurgy. The author also acknowledges and discusses issues of language and the interpretation of terminology used by ironworkers over a period of centuries. A compilation of the many definitions related to iron and steel and their changing meanings is an important component of our survey of the history of the steel- and toolmaking strategies and techniques and the relationship of these traditions to the accomplishments of New England shipsmiths and their offspring, the edge toolmakers who made shipbuilding tools.

The *Hand Tools in History* series is an ongoing project; new information, citations, and definitions are constantly being added as they are discovered or brought to the author's attention. These updates are posted weekly on the museum website and will appear in future editions. All volumes in the *Hand Tools in History* series are or will soon be available in hard copy editions. All other volumes are or will soon be available as bound soft cover editions, are for sale at The Davistown Museum, Liberty Tool Co., local bookstores and museums, or by order from Amazon.com and BookSurge affiliated bookstores.

Table of Contents

INTRODUCTION

Steel- and Toolmaking Strategies and Techniques before 1870 explores ancient and early modern steel- and toolmaking techniques and strategies, including those of early Iron Age, Roman, medieval, and Renaissance metallurgists and toolmakers. Direct process methods for smelting wrought and malleable iron and natural steel dominated ironworking technology from the first dateable appearance of an iron-producing culture at the height of the Bronze Age until the appearance of the blast furnace in western Europe (± 1300). By 1500, the indirect process of decarburizing cast iron to produce wrought and malleable iron and German steel dominated iron- and steelmaking strategies after the transition was made to utilizing the larger and more efficient Stuköfen and blast furnaces. Archaeometallurgists Barraclough (1984), Tylecote (1989), Tweedle (1987), Wertime (1962, 1982), Wayman (2000), and many others have been useful guides for our journey through the labyrinths of ancient metallurgy. The rapid increase in the production of iron and steel after the introduction of blast furnace technology led to the emergence of competitive Renaissance era European trading economies and the consequent exploration, conquest, and settlement of the American continent. The ability of Renaissance era ironmongers to utilize a variety of strategies and techniques to make steel and iron tools and weapons laid the foundation for the later evolution of a robust community of bloomsmiths, shipsmiths, toolmakers, and other ironmongers in colonial America. The colonial finesse at ironmongering played a major role in the success of the American Revolution and the American factory system of manufacturing and toolmaking which followed, subjects explored in volumes 7 and 8 of the *Hand Tools in History* series. This first volume (6) includes an extensive bibliography pertaining to steel- and toolmaking techniques from the early Bronze Age to the beginning of bulk-processed steel production in 1870.

The Davistown Museum exhibition "An Archaeology of Tools" ends with the classic period of the American toolmaker (1827 - 1930) and the rapid expansion in the variety of iron and steel alloys, manufacturing processes, and tool designs that characterize this industrial florescence. As we move back to earlier periods of America's technological history, the tools in early settlers' tool kits were either imported from Europe by immigrants or commercial trading companies or made by blacksmiths, shipsmiths, and edge toolmakers in primitive bog iron bloomeries, fineries, or blacksmith shops. This process naturally raises questions about the origins and metallurgical composition of early tools in our collection that were found in New England not only in archaeological sites but also in workshops, cellars, old factories, and other industrial environments, which are in many cases one or two centuries old.

These questions lead us back to a robust tool manufacturing milieu in continental Europe, which arose in conjunction with the construction of larger and more efficient blast

2

furnaces for smelting iron. Peel back the palimpsest of hand tool production one more layer, and, in the medieval period (600 – 1100 AD), we encounter an already well established division of labor, where bloomsmiths operating direct process bowl and shaft furnaces produced wrought and malleable iron and natural steel in the form of currency bars that were widely traded throughout eastern and central Europe. Blacksmiths and sword cutlers used these currency bars, which consisted of a heterogeneous mix of wrought and malleable iron and natural steel to produce a wide range of artifacts from primitive iron tools and weapons to the highly crafted swords of the Merovingian, Ottonian, and Viking sword cutlers. In turn, these robust migration and medieval period iron-working communities were the descendents of an even more vigorous Roman and Romano-British iron industry whose precedents and evolution are the subject of the first chapters in this volume of the Davistown Museum's *Hand Tools in History* publication series.

It is the legacy of 3800 years of ferrous metallurgy, which lead to the rise of New England's vigorous edge tool industry and, after 1870, to the brief reign of the classic period of American toolmakers.

The following timeline provides an overview of some of the most important events in the history of ferrous metallurgy, topics explored in more detail in all three volumes of historical essays in the six volume *Hand Tools in History* series.

Time Line

Table 1. The following time line provides a brief summary of the chronology of important events pertaining to the Davistown Museum survey of the history of steel- and toolmaking strategies and techniques. Descriptions with an * are excerpted from Barraclough (1984a).

Date	Event
1900 BC	First production of high quality steel edge tools by the Chalybeans from the high quality iron sands of the south shore of the Black Sea.
1200 BC	End of the Bronze Age in the eastern Mediterranean region; steel is probably being produced by the bloomery process.*
800 BC	Carburizing and quenching are being practiced in the Near East.*
800 BC	Beginning of the European Iron Age. Celtic metallurgists begin making natural steel in central and eastern Europe.
650 BC	Widespread trading throughout Europe of iron currency bars, often containing a significant percentage of raw steel
400 BC	Tempered tools and evidence for the 'steeling' of iron from the Near East*
300 BC	The earliest documented use of crucibles for steel production was the smelting of Wootz steel in Muslim communities (Sherby 1995a).
200 BC	Celtic metallurgists begin supplying the Roman Republic with swords made from manganese-laced iron ores mined in Austria (Ancient Noricum).
55 BC	Julius Caesar invades Britain
50 BC	Ancient Noricum is the main center of Roman Empire ironworks. Important iron producing centers are also located in the Black Mountains of France and southern Spain.
43-410 AD	Romans control Britain.
125 AD	Steel is made in China by 'co-fusion'.*
700	High quality pattern-welded swords being produced in the upper Rhine River watershed forges by Merovingian swordsmiths from currency bars smelted in Austria and transported down the Iron Road to the Danube River.
1000	First documented forge used by the Vikings at L'Anse aux Meadows (Newfoundland)
1250-1350	First appearance of blast furnaces in central and northern Europe
+/- 1465	First appearance of blast furnaces in the Forest of Dean (England)
1509	["Natural" (German)] steel made in the Weald [Sussex, England] by fining cast iron*
1601	First record of the cementation process, in Nuremberg*

Date	Event
1607	First shipsmith forge in the American colonies used at Fort St. George, Maine
1613/1617	Cementation process is patented in England.*
1625	First Maine shipsmith, James Phipps, working at Pemaquid
1629-1642	The great migration of Puritans from England brings hundreds of trained shipwrights, shipsmiths, and ironworkers to New England.
1646	First colonial blast furnaces and integrated ironworks are established at Quincy and Saugus, Massachusetts.
1652	James Leonard establishes the first of a series of southeastern Massachusetts colonial era bog iron forges on Two Mile River at Taunton, Massachusetts.
1686	First documented use of the cementation process in England
1703	Joseph Moxon ([1703] 1989) publishes *Mechanick Exercises or the Doctrine of Handy-Works*.
1709	Abraham Darby discovers how to use coke instead of coal to fuel a blast furnace.
1713	First appearance of clandestine steel cementation furnaces in the American colonies
1720	First of the Carver, Massachusetts blast furnaces established at Popes Point
+/- 1720	William Bertram invents manufacture of 'shear steel' on Tyneside.*
1722	René de Réaumur (1722) provides the first detailed European account of malleableizing cast iron.
1742	Benjamin Huntsman adapts the ancient process of crucible steel-production for his watch spring business in Sheffield, England.
1758	John Wilkinson begins the production of engine cylinders made with the use of his recently invented boring machine.
1763-1769	James Watt designs and patents an improved version of the Newcomb atmospheric engine, i.e. the steam engine.
1774	John Wilkinson begins the mass production of engine cylinders used in Watt's steam engine pressure vessels.
1775	Matthew Boulton and James Watt begin mass production of steam engines.
+/-1783	The approximate date when Josiah Underhill began making edge tools in Chester, NH. The Underhill clan continued making edge tools in NH and MA until 1890
1783	James Watt improves the efficiency of the steam engine with introduction of the double-acting engine.
1784	Henry Cort introduces his redesigned reverbatory puddling furnace, allowing the decarburization of cast iron to produce wrought and malleable iron without contact with sulfur containing mineral fuels.

Date	Event
1784	Henry Cort invents and patents grooved rolling mills for producing bar stock and iron rod from wrought and malleable iron.
1802-1807	Henry Maudslay invents and produces 45 different types of machines for mass production of ship's blocks for the British Navy.
1804	Samuel Lucas of Sheffield invents the process of rendering articles of cast iron malleable.
1815-1835	The factory system of using interchangeable parts for clock and gun production begins making its appearance in the United States.
1818	Thomas Blanchard designs a lathe for turning irregular gunstocks.
1828	Adoption of the hot air blast improves blast furnaces
1831	Seth Boyden of Newark, NJ, first produces malleable cast iron commercially in the US.
1832	D. A. Barton begins making axes and edge tools in Rochester, NY.
1832-1853	Joseph Whitworth introduces innovations in precision measurement techniques and a standardized decimal screw thread measuring system.
1835	Steel is first made by the puddling process in Germany.*
1835	The first railroad is established between Boston and Worcester, Massachusetts.
1837	The Collins Axe Company in Collinsville, Connecticut, begins the production of drop-forged axes.
1837	In England, Joseph Nasmyth introduces the steam-powered rotary blowing engine.
1839	William Vickers of Sheffield invents the direct conversion method of making crucible steel without using a converting furnace.
1842	Joseph Nasmyth patents his steam hammer, facilitating the industrial production of heavy equipment, such as railroad locomotives.
1849	Thomas Witherby begins the manufacture of chisels and drawknives in Millbury, MA.
1850	Joseph Dixon invents the graphite crucible used in the American production of cast steel.
1853	John, Charles, and Richard T. Buck form the Buck Brothers Company in Rochester, NY, after emigrating from England and working for D. A. Barton. They later move to Worcester, MA in 1856 and Millbury, MA in 1864.
1856	Gasoline is first distilled at Watertown, Massachusetts.
1856	Bessemer announces his invention of a new bulk process steel-production technique at Cheltenham, England.*
1862	Robert F. Mushet sets up his Titanic Steel and Iron company in the Forest of Dean and begins producing alloy steel with tungsten and titanium.

Date	Event
1863	First successful work on the Siemens open-hearth process*
1865	Significant production of cast steel now ongoing at Pittsburg, Pennsylvania furnaces
1868	R. F. Mushet invents 'Self Hard,' the first commercial alloy steel.*
1874	Tilting band saw is introduced and revolutionizes shipbuilding at Essex, MA.
1879	Sidney Gilchrist Thomas invents basic steelmaking.*
1906	The first electric-arc furnace is installed in Sheffield.*
1913	Brearley invents stainless steel.*
1926	The first high-frequency induction furnace in Sheffield*

*(Barraclough 1984a, 13-4).

Ancient Toolmaking Techniques

The Origins of Metallurgy

The robust English toolmaking industry of the 18th century, its continental predecessors, and unacknowledged siblings, the edge toolmakers of colonial New England, have roots in iron and bronze tool production technologies that can be traced back millennia before the Christian era. A familiarity with the origin and history of metallurgy before 1350 helps to illuminate the tool manufacturing techniques that changed very little between 1350 and 1700. This continuity contrasts sharply with the increasing pace of technological change that characterized tool manufacturing processes after 1700. A review of the origins of metallurgy and the history of edge tool and weapons production provides the historical context for understanding the challenge that the first European settlers in America faced in manufacturing or importing the edge tools so crucial for their survival. For thousands of years before the Industrial Revolution, edge tool production, including weapons, such as swords and knives, was an esoteric, alchemical, and even magical process, the efficiency of which determined the rise and fall of empires.

The terms "Copper Age," "Bronze Age," and "Iron Age" can be misleading. They denote historic eras during which one tool form or another dominated the surviving remnants and relics of a particular culture. In this context, the age of metallurgy – the smelting and casting of metal bearing terrestrial ores – can be traced back to at least 4500 BC, and possibly 6000 BC (Renfrew 1973). All of the many writers on early metallurgy (Percy 1864, Swank 1892, Forbes 1950, Ceram 1956, Smith 1960, Piggott 1965, Plenier 1969, Snodgrass 1980, Wertime 1982, Barraclough 1984a, and Tylecote 1987) concur in noting the rise of a major center of iron and natural steel production (after 2000 BC) in the Caucasus Mountains of northern Turkey and on the shores of the Black Sea, a body of water providing convenient transportation for the metal products of this resource-rich mountain region. Of particular interest is the fact that the Black Sea is lined with self-fluxing iron sand with a magnetite content as high as 80% (Wertime 1982). This sand washes down the rivers from the mountains that lie above the Black Sea on the Turkish coast. The Chalybeans appear to be the first ironmongers to utilize this sand for manufacturing iron and steel implements, circa 1900 BC. Citing Piaskowski (1982), Wertime (1982), indicates that at this time they "were making a high nickel steel by adding the nickel arsenide, chloanthite, consisting of iron, nickel, cobalt, arsenic, and sulfur, to the smelt" (Wertime 1982, 20). Piaskowski (1982) cites classical sources including Xenophon, Euripides, and especially Aristotle, as noting the evolution of the Chalybean Age of Steel at the height of the Bronze Age. Aristotle notes, "that a stone called pyrimachos is thrown" into the furnace during the smelting of iron to help produce the steel tools characteristic of the Chalybean smelting process. The nickel content of this

8

additive or flux (9.44%) helps explain why tools made by the Chalybeans had been frequently mistaken for meteor-iron-derived tools, and also explains their tendency not to rust, a phenomenon also cited by Aristotle. Piaskowski (1982) notes that Chloanthite is a sulfide, also in agreement with the observations of this steel-producing culture made by Euripides.

The iron and steel tools and weapons produced along the edge of the Black Sea were exported to nearby communities for at least a century after 1900 BC. An Assyrian colony was established at Kanes, which continued manufacture of ironware, later supplying the Hittite Empire to the southwest. At no point during the first half of the second millennium BC were sufficient quantities of iron weapons and tools produced to overshadow bronze tool production. In fact, the Anatolian coast was the location of a polymetalic culture, which exploited local resources of copper, silver, and gold, as well as iron. The location of tin deposits essential for the manufacture of bronze tools has not yet been established with certainty. Wertime (1980), Snodgrass (1971, 1980), and others note the extensive trading networks necessary for obtaining tin in distant locations, as well as the possibility that the disruption of the tin trade played a major role in bringing the late Bronze Age to an end. The inability to continue large scale production of bronze weapons combined with longstanding knowledge of iron-smelting techniques to give rise to the Iron Age and its steeled edge tools and weapons after 1200 BC. Though only occurring randomly in the context of bronze tool production, many of the high nickel content iron artifacts noted throughout Egypt and the Near East after 2000 BC may have derived from the Chalybean nickel alloy natural steel. Archaeometallurgists as recent as Tylecote (1976) had previously assumed that these high nickel content iron tools were wrought from meteorites high in both nickel and iron.

The innovative iron-smelting and fining techniques of the residents of northern Turkey followed almost three millennia of copper working and smelting. Wertime (1980), among the foremost of pyrotechnical historians, notes the fact that the use of fire for smelting copper and producing pottery, glass, and concrete from lime is the fundamental element in the evolution of civilizations and their urban environments. The first and foremost products of the Caucasus region, including the Black Sea coastline, were copper implements (tools, utilitarian metalware, and ornamental metalwork) produced as early as 4500 BC. Renfrew (1973) also notes the florescence of a copper-smelting culture in the central Balkans (Vinca – Yugoslavia) as early as 6000 BC. The evolution of a polymetalic culture in the central Balkans may be autochthonous and the first such example of a pyrotechnic culture; its influence on the later Black Sea coastline polymetalic community is unknown and undocumented. Heskel (1980) also notes the rise of an early copper-smelting culture at Tepe Yahya, Iran, with copper awls appearing as early as 4500 BC.

The abundance of copper-containing ores made the Caucasus region a center of metallurgy for over a millennium prior to the appearance of bronze tools. Copper implements and tools have two forms: those produced from the cold hammering of the easily shaped raw copper and those cast from smelted carbonate and oxide ores, such as azurite and malachite. These copper products were produced in the southern Caucasus region and imported in large quantities to Sumer by 3000 BC. At this time, copper metallurgy can be documented as well established in Egypt, China and India. It is also at this date that bronze tools and artifacts begin appearing in large quantities in many cultures. The key transition point in the evolution of a non-pyrotechnic culture to a polymetalic pyrotechnic urban culture occurred when the hammering and shaping of found or mined copper, silver, and gold was supplanted and, in fact, replaced by smelting of copper in furnaces, which reduced or extracted the pure metal from their oxide ores. In the case of copper-smelting, at least in southwest Asia including Anatolia, iron oxide was sometimes used as a flux, producing "bears" of iron slag at the bottom of furnaces, many of which were retrieved and recycled during the Iron Age for the valuable metal they contained. Knowledgeable coppersmiths therefore had early experience with iron. It just took a few centuries or longer to figure out useful applications for the iron slag left at the bottom of the copper-smelting furnaces. The presence of this iron slag helps explain the occasional appearance of iron tools and implements throughout the Bronze Age.

The start of the Bronze Age is marked by the use of arsenic containing ores in the copper-smelting process, which resulted in the first production of a bronze alloy, i.e. arsenic bronze. Arsenic bronze could be hammered, heated, and reforged into edge tools and weapons, which were superior in strength to hammered copper tools. At some point in the late 4th millennium, copper smelters and coppersmiths discovered that if they added tin as an alloy to smelted copper, tin bronze tools were superior to both the arsenic bronze and pure copper tools and implements formerly being produced.

Forbes (1950) provides this historical sketch of the history of metallurgy. Of particular importance is the transition between Stage II, the hammering of native metals, and Stage III, the smelting techniques of polymetalic societies that soon gave rise to experiments with alloys and forging techniques, which greatly expanded tool forms, availability, and quality.

Table 2 Evolution of Metallurgy

I	Native metal as stones [no alterations]
II	Native metal stage ([altered by] hammering cutting etc.) (copper gold silver meteoric iron)

III	Ore stage ([altered] from ore to metal alloys [in a furnace;] composition as primary factor) (lead silver copper antimony tin bronze brass)
IV	iron stage (processing as primary factor) (cast iron wrought iron steel)

(Forbes 1950, 9, Fig. 4)

Initially, the direct reduction of metal from ore was the objective of this industrial activity. Smelters, forge masters, and metalsmiths soon learned that direct process metals production was not their only option. The deliberate production of arsenic bronze, followed by tin bronze, was only the first step in experimenting with alloys and additives that led to the more sophisticated refining processes necessary to produce carburized iron, i.e. steel.

For 3,700 years the strategies and processes for producing steeled edge tools and weapons were based on what Barraclough (1984a) called "rule of thumb" procedures. During this entire period, the critical role of charcoal-derived carbon as a component of steel and the chemistry of fully martinized steel were unknown. The spread of the Iron Age into central and, finally, coastal Europe was the result of the growing intuitive and empirical knowledge of techniques and technologies that could produce steel edge tools and weapons superior to the bronze implements that had been used for thousands of years. The widespread appearance of the blast furnace after 1400 AD in Europe, 2000 years after its first appearance in China, marks the beginning of the emergence of sophisticated pyrotechnical polymetalic empires, well armed and ready to settle and exploit the new-found-lands of North and South America and the West Indies.

The Bronze Age

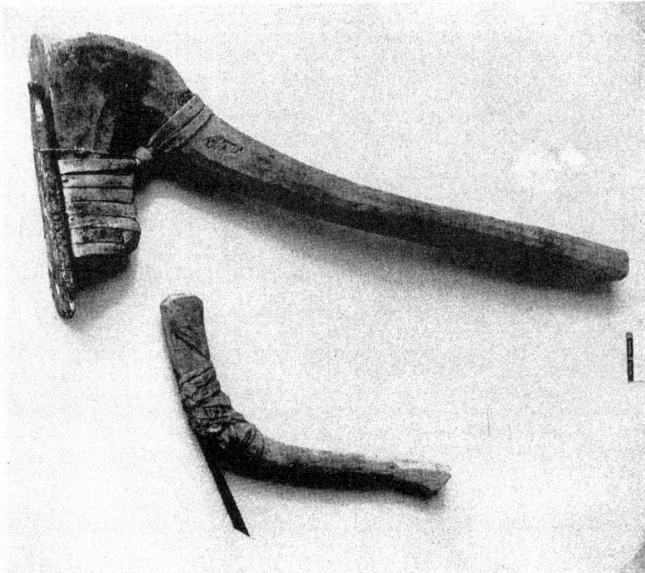

8. Egyptian Copper Adzes, 17th and 18th Dynasties

Figure 1. (Goodman 1964, 18). This Egyptian adz has either a copper or bronze blade with long handles attached to the tool with leather bindings. The adz was not only one of Egypt's most important tools but also the universal ship carpenter's tool until the advent of the water-powered rotary saw (1825) and steam- and electric-powered woodworking machinery (1860, 1890). The evolution of the forms and metallurgy of the adz led to an Industrial Revolution that manufactured exquisite crucible steel adzes, while also rendering obsolete the sailing ships they built.

The use of a technological model to describe the chronology of history can be misleading. This is especially true of prehistoric societies and protohistoric cultures with cuneiform texts that were difficult to interpret. Bronze tools in large quantities began appearing in the Near East and in China at or before 3000 BC. Iron artifacts fashioned from nickel laced meteor iron or from the addition of nickel-containing chloanthite into the iron-smelting process have been found in Egypt and Mesopotamia, dating well before 3000 BC. Fisher (1963) notes the anomalous discovery of non-meteoric forged iron in the Great Pyramid of Greh in Egypt, c. 2900 BC. Other anomalies include the production of malleable cast iron in China after 500 BC and steel-producing bloomeries in south eastern Africa (500 AD).

The occasional appearance of forged iron artifacts during the Copper and Bronze Ages raises intriguing questions about the alchemy of metallurgy. The chemistry of bronze- and iron-smelting was unknown until the very last decade of the 18th century. Yet the art of smelting as a mysterious ritual produced bronze weapons far superior to those made of copper by 3000 BC. There are many questions that remain unanswered. Did the metallurgical magicians of this period also smelt forged iron but discard it as too dull for weapons? Did the art of metallurgy arise at different times and places, as in early Egypt or the central Balkan highlands, only to be lost again and rediscovered? Because of the secrecy surrounding the magic of steel production, did the knowledge of the techniques of producing natural or raw steel arise in some other location, only to return to the Caucasus region around 2000 BC? Did production in India of Wootz steel, an early form of crucible steel, predate iron-smelting in the Caucasus?

The knowledge of casting bronze from copper and tin pre-dates the growing awareness of how to produce steel tools from primitive iron furnaces and bloomeries that arose after 2000 BC, despite the fact that the smelting of bronze is a much more sophisticated process than the smelting of iron. Hammered bronze edge tools and weapons retain a harder and sharper edge than any edge tools and weapons made from wrought or malleable iron. Early metallurgists who smelted bronze probably also knew how to smelt iron; the bronze weapons and tools they were making simply had more durability and usefulness than the same artifacts made from wrought or malleable iron. The length of the Bronze Age, as defined as a period of time characterized by the predominant use of bronze tools, varies among geographical locations. In northern Mesopotamia, a robust steel-producing culture, the Chalybeans, arose at the height of the Bronze Age, circa (1900 BC). Natural steel tools suddenly appeared in large quantities but were not sufficient in their patterns of distribution or their variety of applications to constitute an alternative to the bronze tools commonly used throughout western Asia. After 1200 BC, there was, however, a sudden disruption in the availability of tin, which brought a rapid end to the Bronze Age in the eastern Mediterranean region, including Greece, Cyprus, and Levant. At this date, the florescence of a Chalybean nickel steel-producing culture was ancient history; the dominant forms of the new ferrous metallurgy were wrought and malleable iron and, occasionally, natural steel, and the weapons and edge tools that were derived from their production. In continental Europe, the Bronze Age ended in 750 BC. In Britain, a gradual end came between 500 BC (southern England, including the Thames Valley) and 100 BC (Ireland). Denmark and Iceland continued the use of bronze tools until 100 AD.

One fact stands out with respect to the era of bronze metallurgy: the single most important result was improved weaponry, as with all later developments in metallurgy. Bronze swords and daggers were far superior to copper weapons. The availability of these weapons must have played a major role in the relative ascendancy of one civilization over another in Mesopotamia during the millennium and a half after 3000 BC. Historians may argue that other factors, such as soil fertility, language development, social organization, or religious values in a subsistence economic model of the development of early societies played a role equally important to weaponry. However, in the long run of history the quality and efficient use of weaponry determined the viability and fate of the society more than any other cultural factor.

Weapons were not the only edge tools being produced by early pyrotechnic societies. Hand tools familiar to 18th and 19th century shipwrights have their roots in ancient tool forms, including those made of stone, copper, and bronze. The earliest stone adz was an eolith, which was a found stone edge tool later sharpened by grinding. Many tool forms used by American colonists to build their coasting vessels had forms that can be traced back thousands of years, as, for example, shaft-hole adzes and axes, frame and whip saws, and augers. Early modern strategies for making the steel (e. g. German steel) for edge tools were only a few centuries old in 1608, when Maine's first ship the Virginia was built. The production of blister steel for steeling edge tools was still decades in the future.

Other issues related to metallurgy are also important. The various sources on the history of tools and metallurgy disagree about a diffusion model of history. Did advances in bronze casting start in the Caucasus and gradually advance to China, where bronze casting of religious statues achieved a remarkable degree of sophistication in the second millennium? Or was this art independently developed within China? And how far in the past will future archaeometallurgists be able to trace China's use of cast iron, which Barraclough (1984a) dates back to at least 700 BC?

The labyrinths of the history of metallurgy are filled with many unanswered questions. In a study of the archaeology of tools, did the invention and use of copper and bronze woodworking tools also originate in the Caucasus Mountain region? When the metal products of the Caucasus region were transported by sea to other locations, were the ships used built by stone adzes or bronze adzes, or did similar tools forms arrive spontaneously in several locations? What impact did the use of bronze edge tools have on the woodworkers of pre- and protohistory? Did the spread of bronze woodworking edge tools accompany the spread of bronze weapons? Was there a time lag, with stone edge tools lingering in use after the appearance of bronze weaponry? Were bronze edge tools superior to the ground stone axes and adzes used by the first shipbuilders, either Mesopotamian or northern European?

The following illustrations are of typical Bronze Age edge tools. In Figure 3, the socket-hole tools illustrate an anomalous tool form that lingered until the beginning of the Iron Age in northern Europe without further development. In many cases far older than the socket-hole bronze edge tools, the earlier shaft-hole edge tools in Figure 2 are prototypes for modern edge tools. In several cases [Figure 3: (c), (e), and (h)], these ancient tool forms are similar to modern American axes.

The presence of these anomalous tools in the archeological palimpsests (layers) of the accidental durable remnants of human culture reminds us that the historical evolution of metallurgy is not a straightforward series of events in time. The evolution of the human use of stone, then metal, chemical, electronic, and nuclear tools occurs in the form of the construction of interwoven and ecologically symbiotic labyrinths, i.e. human ecosystems, which overlap each other in time and space, thwarting easy historical narration. The essence of human ecology is the pyrotechnic, polymetalic manufacture of tools to form anthropogenic ecosystems, which are then superimposed upon naturally occurring ecosystems. The study of hand tools in history inexorably leads to a confrontation with

14

the impact of tool-wielding human culture on a biosphere with many limiting factors. The evolution of the art and science of ferrous metallurgy from prehistory to the modern era brings us to the dawn of the age of biocatastrophe. The evolution of the Bronze Age tool forms illustrated by Goodman (1964) is a reminder of the longevity of the human effort to conquer and exploit both natural ecosystems and the neighboring communities who were already competing for scarce natural resources. It is in this context that this volume explores the roots and evolution of the art, and later, the science of ferrous metallurgy.

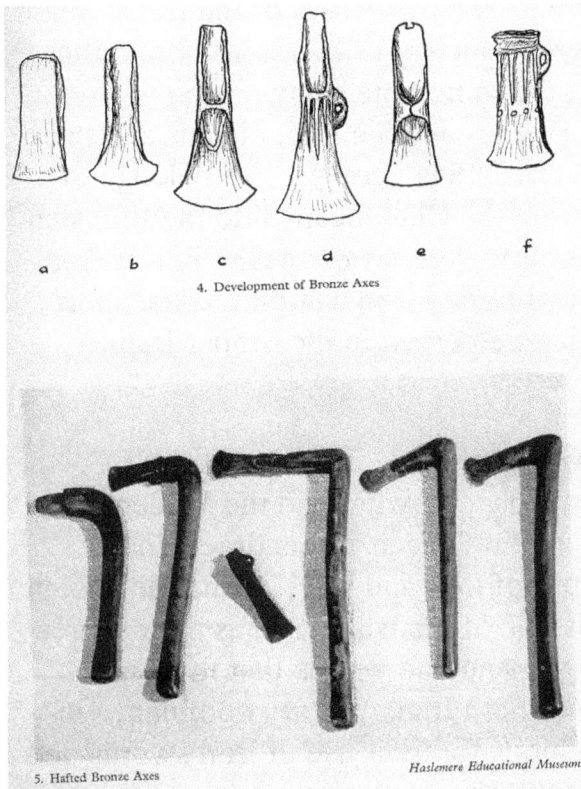

4. Development of Bronze Axes

5. Hafted Bronze Axes

Haslemere Educational Museum

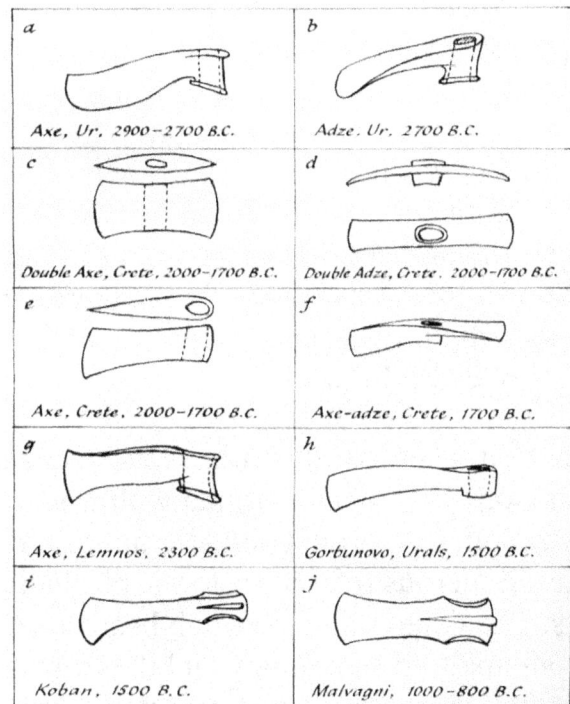

a	b
Axe, Ur, 2900–2700 B.C.	Adze, Ur, 2700 B.C.
c	d
Double Axe, Crete, 2000–1700 B.C.	Double Adze, Crete, 2000–1700 B.C.
e	f
Axe, Crete, 2000–1700 B.C.	Axe-adze, Crete, 1700 B.C.
g	h
Axe, Lemnos, 2300 B.C.	Gorbunovo, Urals, 1500 B.C.
i	j
Koban, 1500 B.C.	Malvagni, 1000–800 B.C.

9. Bronze Shaft-hole Axes and Adzes: (a and b) British Museum; (c, d, e, and f) Heraklion Museum, Crete; (g) National Museum, Athens; (h and i) Material and Researches in 'Archeology', No. 35, Moscow, 1953; (j) Museo Nazionale, Syracuse, Sicily

Figure 2. Left (Goodman 1964, 15). Goodman illustrates an interesting cul-de-sac in edge tool design: northern Europeans clung to an obsolete model of socket-hole bronze edge tools throughout the European Bronze Age, requiring the hafting of the tool with a knee-shaped handle, as illustrated in the lower figure.

Figure 3. Right (Goodman 1964, 20). The modern design of the shaft-hole ax or adz first appears in the tool kits of Sumerians circa 3000 BC. The Goodman illustrations depict shaft-hole edge tools used throughout Mesopotamia, the Mediterranean, and the Russian Steppes. These edge tools were much more efficient than the socket-hole edge tools used in Barbarian Europe during its Bronze Age. These illustrations also show the wide variety and relatively modern design of Bronze Age edge tools during this period. The ax illustrated in (h) is a type encountered in Rome a millennium later. The axe-adze from Crete illustrated in (f) is reminiscent of the combination tools characterizing the tool kits of Vikings two millennia in the future. In both cases, these tools were made of iron, but their prototypes are illustrated here. Why European Bronze Age toolmakers never utilized the design of the earlier shaft-hole style stone tools remains an unsolved archaeological mystery.

15

The Evolution of Iron Metallurgy

The origins of iron-smelting appear to have the same geographical Indo-European roots as copper and bronze metallurgy, the Caucasus region. Snodgrass (1980), Wertime (1982), Plenier (1969), and other writers note the florescence of a full-fledged Iron Age characterized by steeled edge tools and weapons centered first at Cyprus in the mid-13th century BC with major centers at Crete, the Peloponnesus and Athens by 1200 BC. The earlier isolated ironworking technology associated with the iron sands of the Black Sea is the probable source of this florescence of a pyrotechnic society utilizing iron ore rather than copper and tin as its major source material. At the same time ironworking technology was working south into the land of the Hittites and the Phoenicians and then west to Cyprus, possibly with the help of the mysterious "Sea Peoples," knowledge of ironworking and steeling techniques spread north from the Chalybeans into the Eurasian steppes of southern Russia, resulting in the Scythian Iron Age. Ironworking in this area included a robust tradition of animal art work, conical footed iron cauldrons, and short iron swords and daggers, prototypes of forms that later emerged in the timber frame burials of central Europe.

Bronze edge tools still dominated Mycenaean culture until its demise circa 1100 BC, when Dorians, probably armed with iron or steel swords, overwhelmed the Mycenaean civilization. Iron production also appears on Crete and in Greece at this time; Homer makes numerous references to the production and use of iron and steel during the Trojan Wars. The Greeks cite the Chalybeans on the south side of the Black Sea as their source of iron-smelting technology and the iron swords, spears and horse bits that followed. With the fall of the Hittites in eastern Turkey and northern Iran, military dominance in northern Mesopotamia evolved to the Assyrians, famous for their use of iron swords and especially, iron chariots, probably the most important technological innovation of the Iron Age. One of the enduring mysteries of Mediterranean protohistory is the extent to which iron and steel weapons were used by the invading "Sea Peoples" during the great migrations of 1250 - 1150 BC. These migrations by numerous small bands of warriors from communities as diverse as Libya, Anatolia, Greece, and possibly more northern locations put an end to late Bronze Age civilizations from Egypt to Troy.

After 1000 BC, the Iron Age spread rapidly into southern and central Italy and then into Austria and Germany by 700 BC. Knowledge of ironworking technology in central Europe after 700 BC appears to have roots both in the culture of the Scythian ironmongers of the Russian Steppes and in the sophisticated ironworking centers of southern Italy and Etruria. The presence of a vigorous Celtic ironworking industry in southern Britain by 500 BC is clearly demonstrated by the large number of iron tools on display at the Museum of London, many of which pre-date the arrival of Roman invaders

(55 BC), who were motivated, in part, by the existence of this well established industry. Phoenician traders and ironworkers also spread knowledge of iron-smelting and steeling techniques to coastal France and especially to Spain during the early years of the last millennium BC (Plenier 1980).

Iron-smelting technology in India, which evolved into a sophisticated crucible steelmaking (Wootz steel) capability, may have also originated in the Caucasus. However, Wootz steel had its limitations, as it was produced only in 1 to 3 kg batches and required mixing into the crucible just the right amount of pure wrought iron and a carboniferous material, such as granulated charcoal. Some recent commentators (Smith 1960, Williams 1977a) suggest that Wootz steel production was based on the fusion of melted cast iron and wrought iron in these crucibles. The exact production methods of Wootz steel discs, which were high in iron carbide and possibly subject to very slow cooling (Williams 1977a), probably varied from region to region (India, Persia) and may never be understood. Wootz steel, in the form of circular discs, was later traded to Damascus, where the famous Damascus steel swords were produced, possibly as early as 1500 BC. The forge welding of Damascus steel swords involved the same technique of pattern-welding thin sheets of crucible steel and malleable or wrought sheet iron, producing malleable steel swords of amazing durability and quality. The Damascus sword was of a higher quality than any weapon produced from malleable iron, natural steel, forged steel, or case-hardened steel. Archaeometallurgical analysis indicates that many swords were "refined" as many as 25 or 50 times (Barraclough 1984a). While Barraclough is referring here to pattern-welded weapons made in China, circa 77 AD - 189 AD, the same pattern-welding techniques were universally used throughout the Near East and central Europe to make the highest quality swords of the early Iron Age. No examples of swords made from Wootz steel have been documented in Europe prior to the Viking era. Pattern-welded swords were, however, forged almost as soon as the first iron and natural steel tools were produced. In the area of the Caucasus this may have been as early as 2000 BC. The Damascus steel industry was a component of the spread of iron-smelting technology to northern Palestine, which was well established over a broad area by 1250 BC. In the Bible, extensive references to iron (90) and steel production (5) illustrate the spread of this technological process.

As noted, iron-smelting technology originated in the milieu of copper and bronze edge tool and weapons production but may have been discounted or ignored due to the superiority of bronze edge tools and weaponry over duller wrought iron implements. Practical methods for providing large quantities of "steeled" or natural steel edge tools may have been lacking. Whatever the case, when production of iron tools and implements became widespread sometime after 1200 BC, the result was a rapidly expanding technological revolution. This revolution was the result of the growing knowledge of two alternative strategies for producing steel edge tools and weapons. The first was the realization that smelted iron could be made into "natural" steel by halting the decarburization process prior to a bloom of iron becoming low-carbon wrought iron. This was done by operating a shaft furnace or even a bowl furnace at a higher temperature by lowering the angle of the tuyère and increasing the ratio of carboniferous fuel to iron ore. Wertime (1980) notes that the control of ancient shaft furnaces was difficult; carbon re-absorption often resulted in the production of cast iron (higher carbon content resulted in lower melting temperature). The many examples of cast iron artifacts that he cites suggests that cast iron and natural steel with widely variable carbon contents were two well known byproducts of the attempt to smelt low carbon wrought iron.

The second method of making steel was based on an awareness that inserting an iron tool into a charcoal fire for long periods of time in a manner that excluded contact with oxidizing flames resulted in the absorption of carbon and then the formation of a steel edge on the tool being forge welded. The cutting edge of the tool was carburized by submergence in the charcoal, then hammered and again reheated. The result was the forge welding of a steely cutting edge on the tool. Sword and knife cutlers, bill hook and scythe smiths, and woodworking edge toolmakers had no knowledge of the importance of carbon in this forge welding process. These artisans must have been able to differentiate tools made with low carbon malleable iron (< 0.2% carbon content) from the more desirable, but more difficult to make, tools made from iron with a higher carbon content ($\pm$ 0.4%), which were stronger and more durable, if less ductile, than those with the lower carbon content. Nearly pure wrought iron tools (< 0.08% cc) would not have been a desirable product due to their lack of strength and excessive ductility. Only the feel of the tool after forge welding would tell the smith if it was suitable for an appropriate function.

Forge welding is a toolmaking strategy that has endured in various forms throughout history and is still utilized to produce many tool forms. Blacksmiths of the early Iron Age learned that they could make not only forged iron tools with a slightly higher carbon content than wrought iron artifacts, but primitive natural steel tools by these methods. To further improve the quality, the natural steel or carburized steel implement had to be reheated and hammered to expel slag inclusions. Some blacksmiths must have noted that rapid cooling (quenching) of a minority of the tools they were forging produced extremely hard and brittle edge tools. This brittleness could be relieved by tempering, i.e. reheating the tool at a temperature below its critical (melting) point, which alleviated the brittleness and produced durable yet malleable edge tools, especially swords and knives. The repeated, but not common, appearance of fully martinized steel tools in the early Iron Age indicate that at least some toolmakers, including sword cutlers, had learned how to forge high carbon, heat treated (quenched and tempered) natural steel tools and weapons.

The one tool at a time recarburization and forge welding of the outer iron surface of the tool in a charcoal hearth was time-consuming and the danger of oxidizing the carbon

18

content by burning the steel was an ever present problem. The key to forging steel edge tools was to prevent surface contact with combustion gases. Primitive methods to protect the steel from being oxidized included dipping the iron tool in pig fat and wrapping it in goat skin prior to submerging it in the charcoal fire, or variations of covering it with leather, mud, and clay. The fundamental principal was "enclosure" – keeping the wrought or malleable iron implement that was being carburized isolated from combustion gases. The basic principle of early forging techniques was thus similar to that employed in making Wootz, Damascus or Toledo steel in crucibles, where the iron was reheated in direct contact with carboniferous materials in crucibles, which protected the tool from combustion gasses.

A third primitive method for making edge tools and weapons was case hardening, a variation of carburizing and then hand forging steel tools and weapons. A lump of wrought or malleable iron from the bloom was pounded into thin sheets of iron, which were then buried in a bed of charcoal, possibly even a desert campfire. After hours of baking, the outer layers of the sheet iron would be carburized sufficiently to become steel. These sheets of steel could be piled and hammered together and interspersed with sheets of wrought or malleable iron. These alternating sheets of steel and iron could be forge welded into the pattern-welded swords and weapons of the early Iron Age. Victory in warfare often went to the warriors with the best blacksmiths who produced the highest quality weapons by the tedious carburization of wrought or malleable iron or the even more difficult pattern-welding of sheet steel and sheet iron. The intuitive knowledge of hardening by sudden quenching followed by tempering to relieve brittleness was the secret to successful production of the steel edge tools superior in quality to hammered bronze weapons or the vast majority of malleable iron (low carbon natural steel) tools lacking a sufficient carbon content (0.5%) to be successfully quenched and tempered. By 700 BC, knowledge of how to create high quality edge and horticultural tools by quenching and tempering as well as from malleableized cast iron was well established in China, thousands of miles from the locus of the birth of the Iron Age along the shores of the Black and Mediterranean seas.

The Chinese Iron Age

In his classic *Blister Steel: The Birth of an Industry*, Barraclough (1984a) provides a sketch of the many strategies that the Chinese used to produce steel. Following the pioneering research of Needham (1958), who wrote *The Development of Iron and Steel Technology in China*, Barraclough (1984a) notes that the rise of a robust cast iron industry in China can be dated at least as early as 700 BC and appears to be nearly concurrent with the use of the direct process bloomery furnace in Europe to produce iron and steel. There is no readily identifiable period of bloomery iron and steel production that can, as yet, be identified as preceding the rise of cast iron production in China. Perhaps the most significant evidence for the sophisticated steel-producing capacity in China at this time is the existence of a white cast iron adz with a decarburized cutting edge. The cutting edge of this adz has been carefully heat-treated, illustrating a sophisticated ability at malleableizing cast iron. Barraclough (1984a) notes that, by the Menchuis era (4th century BC), there is ample evidence of the widespread production and use of malleableized horticultural tools throughout China. Of particular interest is the fact that the process for malleableizing cast iron apparently went out of use in China around approximately 700 AD. A millennium passed before the French metallurgist R. A. F. de Réaumur (1722) wrote about this process, already possibly known to knowledgeable Celtic metallurgists. In the ethnocentric world of American metallurgy, Seth Boyden is considered to have been the inventor of malleableizing cast iron in the 4th decade of the 19th century.

Barraclough (1984a) describes the fully developed Chinese version of the blast furnace, which was widely used during the Han Dynasty (202 BC – 220 AD). This relatively small furnace with a circular hearth was typically 13 feet in height, 8 feet in diameter, and was charcoal-fired, producing approximately a half ton of cast iron per day. Coal was also used as a fuel source but at a later date. Enough specimens have been recovered from Chinese gravesites to illustrate that, by the third century BC, a wide diversity of techniques existed for producing swords, including sophisticated pattern-welded swords. Barraclough (1984a) notes in particular that the steel produced in China was made by the charcoal fining of pig iron by a process very similar to that used in Germany and Austria (the Styrian process also known as the continental process) fifteen hundred years or more in the future. Barraclough (1984a), after Needham (1958), also notes widespread production of wrought iron as well as the first documented evidence of what is now known as the Brescian process, which was in evidence at least as early as 125 AD. Several variations of producing steel by this process, the submergence of wrought iron in liquid cast iron, can be dated as early as 1116 AD from written sources, which note:

> Now for making steel, they take bars of soft iron and fold them up in coils, inserting pieces of cast iron between the layers. Then they seal up the furnace with clay and heat it. (Barraclough 1984a, 32)

Barraclough (1984a) describes this as a solid-diffusion method, which was widely practiced at this time. The resultant heated mass of raw steel was then forged into a more coherent mass and then subject to further forging and heat treatments to produce some of the fine steel swords that can also be dated from this period. Barraclough (1984a) also quotes Needham's citation of a 1637 manuscript describing the technique of fusion

> The wrought iron is beaten into thin plates or scales as wide as a finger and rather over an inch and a half long. These are all wrapped within wrought iron sheets and tightly pressed down by cast iron pieces placed on top. The whole furnace is then covered over with mud matted with worn out straw sandals. The bottom of the pile is daubed with mud as well. Large furnace piston bellows are then set to work and when the fire has risen to a sufficient heat the cast iron comes to its transformation first, and, dripping and soaking, penetrates into the wrought iron. When the two are united with each other they are taken out and forged; afterwards they are again heated and hammered. This is many times repeated. The product is usually called "lump steel" or "irrigated steel". (Barraclough 1984a, 33)

Barraclough notes that this method anticipates later western methods but omits the ingot casting stage. The fact that this particular strategy for making steel was in use in China at this late date (the manuscript is dated 1637) and follows almost two millennia of use of variations of the Brescian method suggests that there were a wide variety of steelmaking options available to continental European, English, and colonial American immigrant ironworkers in the 17th century. We now think of blister steel production via the cementation furnace as the source of our early steeled edge tools. The use of the cementation furnace is a relatively late development in steel-producing strategies. The many and varied older options for making steel were well known at this time.

The European Iron Age: Halstadt 750 - 450 BC
(The German spelling is Hallstatt)

One of the curiosities of the Iron Age in Europe is that it did not spread, as might be expected, from Greece to Rome to southern Europe and thence north but from the east out of the vast Russian Steppes of the Scythian horsemen with their iron swords and animal art, probably following the same pathway as the bronze swords that preceded them (2000 BC). The earliest examples of iron tools in continental Europe are found in Czechoslovakia wagon graves well to the east of Halstadt, Austria.

The rise of an iron-smelting culture in Europe is time- and site-specific. The first evidence of widespread use of iron tools and implements comes from the vast graveyards (in excess of 2,500 graves) of the later periods of Halstadt culture. During the late Bronze Age (1200 to 800 BC), Halstadt graves contained a vast array of bronze tools, many similar to those in the tool kits of the Scythians to the east. After 800 BC, iron technology was introduced into proto-Celtic Europe and iron tools began appearing in Halstadt grave sites. A flourishing salt trade with southern Europe and the Mediterranean helped support a robust culture with elaborate hill fort burials, some containing four wheeled funerary wagons, and including rich hoards of tools, gold metalwork, and other cultural artifacts. The Halstadt culture extended in the east to Bohemia and Czechoslovakia and thence westward to Switzerland and eastern France. The culture was named after the salt mine town of Halstadt in central Austria, its most important commercial center. The characteristic tool of the later Halstadt culture is a long iron sword, which first appeared in the Czech wagon graves replacing the decorated long bronze swords of the earlier Halstadt culture. Iron edge tools may have been used to the east to construct the larger plank-built burial chambers that contained so many of the iron tools recovered from this culture. A gradual westward shift of Halstadt culture led to the establishment of new trading routes down the Rhine River with connections to the Rhone River, which provided access to the recently established Greek trading center at Marseilles (circa 600 BC).

A fundamental question remains unresolved with respect to Halstadt iron production. Etruria, located to the south of Austria in central Italy, possessed iron tools and artifacts as early as 1100 BC. Yet Halstadt, with its extensive mineral resources and its rich manganese containing iron ore was not an iron tool production center until 800 AD. Did it also import wrought iron or iron tools from the south in trade for salt, its inhabitants also having learned about ferrous metallurgy from Scythian predecessors to the east? Celtic Halstadt's metallurgical skills may have had a second source in the south, arriving via Etruria from Aegean and Mediterranean sources as a result of its Etrurian trade networks.

The European Iron Age: La Téne 450 - 50 BC

La Téne represents the next stage of Celtic culture in central Europe. La Téne was a village and ritual site on the edge of Lake Neuchâtel in Switzerland. Discovered in 1857, La Téne, lying west of Halstadt, was rich in archaeological artifacts, indicating a fusion of styles of the earlier Halstadt chiefdoms to the east with Greek and Etrurian decorative styles to the south. La Téne reflects the gradual movement of Celtic influence to the west, reaching the Atlantic coast of France by 200 BC. La Téne styles in decorated metalwork, tools, and pottery appeared in Spain, throughout Gaul, and eventually in Britain after 100 BC. La Téne burials contain more warrior-related artifacts than the more peaceful Halstadt; these included iron-axle chariots in aristocratic warrior burials. La Téne metalwork is characterized by distinctive curvilinear motifs. Its iron swords and daggers were shorter than those made by Halstadt ironmongers but with greater decorative qualities. The La Téne culture obliterated Halstadt styles and forms. It spread south into Italy and to Britain, though why and how it spread remains unknown. One of the mysteries of the early Iron Age in Gaul, and especially southern and western Britain is that a robust ironworking industry may have preceded La Téne-related stylistic elements by several hundred years, raising the probability of penetration by Halstadt ferrous metallurgy technology as early as 500 BC, especially in Britain.

A second mystery is the relationship between the relatively rich resources of the Sussex Weald, south of London, with its manganese-laced siderite ores, and the Roman invaders who took over the iron industry of the Weald from the indigenous Celtic ironmongers who had worked its ore deposits for centuries. Perhaps the Romans integrated this indigenous iron-smelting community in their efforts to utilize the same resources for arming the legions of Roman warriors who were fighting or would soon fight the Gauls. The Sussex countryside is littered with remains of Roman bloomeries, usually located along streams directly adjacent to ore deposits, and always marked by the tell-tale slag deposits of the smelting process.

By the beginning of the La Téne era, three different strategies for making steel were well established. The primary strategy was natural steel production from direct process bloomeries, which resulted from the finesse of the iron smelter's control of the reducing process. As noted, changes in the fuel to ore ratio, combined with the use of manganese-laced ores, facilitated the occasional production of raw steel blooms, which had to be carefully reforged into serviceable natural steel tools. In most cases, steel bits were forge welded onto iron shafts. A second strategy was the production of sheet iron, which was then carburized by the case hardening process, i.e. the submergence of the sheet iron in a charcoal fire. Sometimes this sheet steel was then repiled and reforged as steel bar stock before being used to steel edge tools or for pattern-welding steel and iron swords. A third strategy for producing natural steel edge tools involved the carburization of wrought or malleable iron by the time-consuming process of submerging individual tools in the charcoal fire for the purpose of the carburization of the cutting edge of the tool being forged. An alternative process was case hardening an edge tool by submerging it in a charcoal fire or fire pit. Only a few edge tools could be produced at one time by these tedious processes.

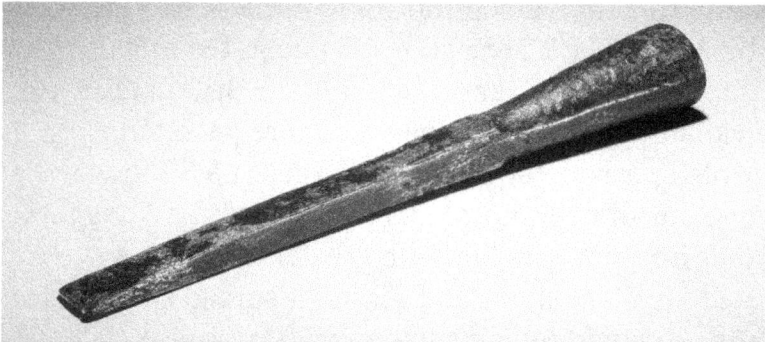

Figure 4. This is an excellent example of a Romano-British, direct process produced, natural steel edge tool circa 200 AD. Note its similarity with the colonial era edge tool found in southeastern Massachusetts illustrated in Figure 5. © The Trustees of The British Museum.

Figure 5. This American colonial era direct process natural steel socket chisel c. 1650 – 1700 is similar to edge tools that would have been made by the direct process at any time from the beginning of the Iron Age to the 19th century. 9 ¼" long, in the Davistown Museum MII collection ID# 41907T1.

Bloomeries specifically designed to produce raw steel probably appeared sometime between the late Halstadt and early La Téne periods. Tylecote (1987), Pleiner (1962), and others note the presence of "currency bars" at locations throughout continental Europe in the early Iron Age. These portable bars of iron often contained significant quantities of raw steel, that is, they had a highly heterogeneous carbon content. Whether they were accidentally or deliberately smelted initially with a high carbon content, currency bars were the raw material used to forge sheet iron and steel from thin bar stock made from these bars. The primary function of finers, blacksmiths, sword cutlers, and edge toolmakers was to create iron and steel tools and weapons from these currency bars, which were traded throughout continental Europe. Before 1300, most raw steel produced by these tedious processes was natural steel derived from direct process smelting furnaces. Most edge tool and sword production using raw natural steel or pattern-welded iron and steel resulted in the ubiquitous production of the undistinguished "iron" swords and tools found in most museum collections today. Two exceptions to the generic use of direct process raw steel in the centuries before 1300 should be noted. The first is the possible, but currently undocumented, use of imported Wootz steel. The second is the recent discovery that the relatively slag-free armor produced by Romano-British blacksmiths may have been derived from the continental method of decarburizing cast iron, a topic explored in later chapters. The artful production of high quality Roman or Merovingian swords, few of which survive today, reflect the exceptional finesse of a few sword cutlers usually associated with royalty.

Iron Age in Romano Britain and the Roman Empire 500 BC - 400 AD

Unlike the beginnings of the European Iron Age, the widespread use of iron tools in Britain cannot be pinned down to a specific time and location. Iron tools, weapons, and other implements began appearing in British archaeological sites dated as early as 500 BC. By 100 BC, the Iron Age appeared to be well established in England but not in Ireland, where use of bronze tools lingered well into the era of the Roman occupation of Britain (43 AD).

Caesar led the first Roman expedition to Britain in 55 BC, then again in 54 BC. During the preceding five centuries, Britain had been gradually settled by bands of Celtic farmers from Gaul. The gradual nature of their settlement may be reflected in the sporadic appearance of iron tools and weapons in Britain in the same period. By the time Caesar encountered Celtic Catuvellauni at Britain's most prosperous city fort at Colchester (54 BC), he found numerous well armed soldiers equipped with iron swords and horse-drawn chariots.

Whatever remnants remained of a Bronze Age culture in Britain were quickly obliterated by the Roman invasion of Britain at Rusborough (AD 43). The 25,000 Roman troops brought iron implements of every description, including natural steel and forged steel edge tools and weapons. This rich legacy of the British Iron Age during the Roman occupation is well illustrated in the many museum collections in England, most notably in the British Museum in London, the Museum of London, and regional museums throughout Britain. As Goodman (1964) illustrates, prototypes for 18th and 19th century American hand and edge tools abound in these collections, in some cases unchanged until the 20th century, as with the early forms of scissors, which evolved into the contemporary form of sheep shears.

Figure 6. Group of tang-type chisels, Romano Britain c. 200 AD © Museum of London.

The exhibition of tang-type chisels at the Museum of London (Figure 6) is of particular interest, illustrating the longevity of basic tool forms. Forged in Roman Britain c. 200 AD and recovered near London, these tang-type chisels were probably used by Thames shipwrights. They have the same basic form as imported English tang-type and socket-type chisels of the late 18th and early 19th century, which are often encountered in New England tool chests and collections.

It is at this point in time that a diffusion theory of technology becomes inadequate. Roman iron and steel had three sources, the most import of which was Noricum, the Roman province that was also the location of the Halstadt culture with its manganese-rich iron so useful in the production of natural steel. Important iron and steel-producing centers in Noricum included the trading centers of Magdaleneberg in southern Austria, Huttenberg to the north, and Linz, at the end of the "Iron Road" linking Erzberg (Iron Mountain) to the Danube River. Spanish mines and metallurgists supplemented the famous natural steel tools made from the spathic ores of the Erzberg in Austria, also supplying iron and steel tools and weapons to the Roman Republic. Contemporary literature discussing specific sources of Roman iron and steel are meager. More well documented are the extensive Roman bloomeries of the Sussex Weald (Cleere 1985), where manganese-laced siderite ores with a similar chemistry to those in Noricum facilitated *in situ* production of Roman malleable iron and steel tools and weapons.

The sacking of Rome by the Gauls in 390 BC may have been due to the gradual improvements in the metallurgy of Gallic swords, which may have increased in quality over several centuries and been made by the same Celtic metallurgists in Noricum who later supplied the Roman Republic, and especially the Roman Empire, with sheet iron and natural steel for their pattern-welded swords. The Etruscans, who controlled northern Italy and the Rome area until 500 BC, also had a large repertory of iron tools and weaponry. As with the origins of the Etruscans themselves, the sources of their iron technology are unknown. It is possible that their knowledge of iron-smelting diffused from the north in Gaul, where there may have been occasional iron production even before Halstadt, possibly as early as 900 BC. Did the Etruscans, who overthrew the Romans in 500 BC, then bring their knowledge of iron-smelting to Rome and central Italy before it filtered in from Greece? Did the knowledge of iron-smelting reach Spain earlier than Italy, courtesy of Phoenician traders and ironmongers who were another source of knowledge of the secret alchemy of ferrous metallurgy? Or did Etruscan iron metallurgy derive from earlier Iron Age activity in Greece, Crete, and Cyprus, currently the conventional viewpoint (Snodgrass 1980, Wertime 1980)?

There are many unanswered questions about Roman ferrous metallurgy. Was the later defeat of the Gauls by the Romans (55 BC), allegedly with superior steel swords, the result of obtaining superior Spanish steel *after* the first successful Gallic invasion of 390 BC? What role did Spanish steel swords play in the defeat of Carthage? Whatever the earlier role of Spanish steel in the rise of the Roman Republic, by the onset of the Roman Empire (44 BC) Roman sword cutlers had perfected the art of sword- and armor-making with the help of the manganese-rich spathic ores from Noricum, which may not have been available in Spain. Again, the labyrinths of the ancient history of ferrous metallurgy lead to many unanswered questions. We don't know if the products of extensive networks of Roman bloomeries in the Sussex Weald were used only to supply Roman swordsmiths and toolmakers, including woodworking and horticultural edge toolmakers, who worked in Britain. Nor do we know the extent to which siderite ores from the Sussex Weald supplemented natural steel produced in Noricum and Spain and used by Roman sword cutlers already well established in the upper Rhine River watershed.

Caesar was assassinated in 44 BC, marking the end of the Roman Republic, the beginning of the Roman Empire, and the ascendancy of an Iron Age that used many of the basic woodworking tools still found in 19th century woodworkers' tool chests. These include cast iron jack and jointer planes, socket-type and tang-type chisels and gouges, axes, adzes, hammers, pry bars, blacksmiths' tongs, and other iron tools (Goodman 1964). The boat shop adz hammer at the British Museum (Figure 7) could easily be mistaken for a late 18th or early 19th century product of a New England toolmaker working in a boatbuilding community; only the sharply curved blade differentiates this tool from a common coopers' adz.

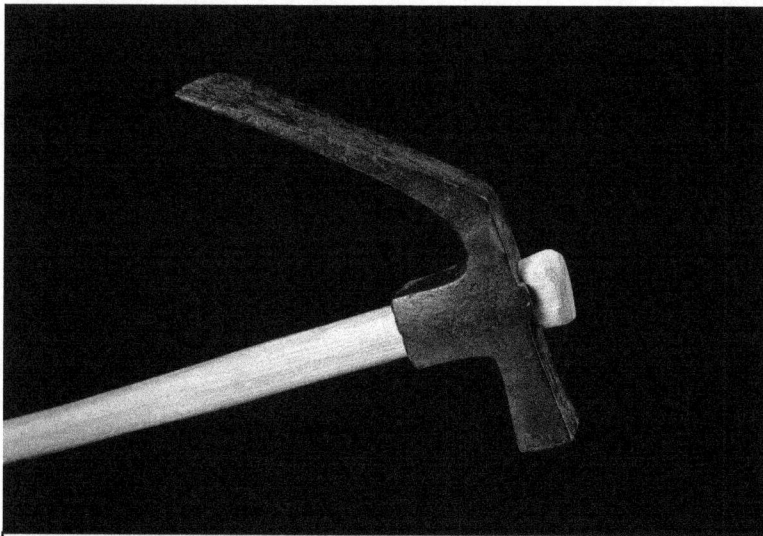

Figure 7. Boat shop adz hammer, malleable iron, with a cutting edge steeled by additional forging. © The Trustees of The British Museum.

During this period, iron production in many locations, such as Sussex, Carinthia, and Spain, was underway on a large scale. A diversity of steelmaking strategies, most of which were derived from direct process bloom-smelting, had been in use for centuries. By the time of the Roman Empire, the hegemony of an Iron Age culture with steel swords, natural steel edge tools, and iron chariot and wagon axles was established from the furthest reaches of Mesopotamia to the northernmost regions of Europe and the fringes of the British Isles.

The Growth in the Diversity of Steelmaking Strategies

A uniform group of steel- and toolmaking strategies and techniques had evolved from the early Iron Age and would remain unchanged until the appearance of the blast furnace in continental Europe after 1350. Hundreds of often finely crafted low carbon natural steel ($\pm$ 0.2 – 0.5% cc, e.g. malleable iron) ax heads, often containing slag inclusions, abound in private and museum collections throughout Europe. Interspersed in these collections are the rare examples of finely crafted, fully martinized, steel edge tools, including swords and axes, which suggest that an elite group of sword cutlers and edge toolmakers produced weapons of exceptional quality for royalty and military leaders, in contrast to the mass-produced swords and knives first forged for the tens of thousands of Roman soldiers who occupied northern Europe and Britain and later for migration and medieval period warriors and woodworkers.

While most of these tools were made from direct process-derived wrought and malleable iron and/or natural steel and utilized the steelmaking strategies already noted, the high quality of the steel in some swords and edge tools raises the possibility of more complex strategies to produce relatively slag-free steel tools and armor. Natural steel is relatively high in slag inclusions. To obtain relatively slag-free steel, sophisticated forging and refining techniques of direct process-produced natural steel or carburized sheet and bar iron are essential. Recent analysis of Romano-British armor (Fulford 2005) suggests the possibility that selective decarburization of direct process bloomery-derived cast iron low in slag by Romano-British fineries resulted in the production of relatively slag-free steel for armor.

> Metallographic examination of several different types of Roman ferrous armor from northern Britain dating between the late first and early third century has revealed a complexity and variety of structures. (Fulford 2005, abstract)

Specimens examined included five cold- and warm-hardened (malleable) iron, one medium carbon iron, one surface carburized specimen, and four pattern-welded sheet iron and steel specimens. The significant characteristic of all specimens was the relative absence of slag inclusions characteristic of direct process bloomery-derived wrought and malleable iron and natural steel. These specimens imply extensive further refining of bloomery-derived billets. As precursors of the later perfection of German steel from decarburized cast iron, these specimens raise the intriguing question of the early roots of the forging of relatively slag-free malleable iron and carbon steel by this technique. It is possible that Romano British bloomsmiths specifically smelted cast iron billets from direct process bloomeries before the appearance of larger, more efficient blast furnaces, such as that at Lapphyttan, Sweden $\pm$ 1250 AD. Such billets would have been relatively slag-free in comparison to bloomery-derived wrought and malleable iron or natural steel.

The deliberate decarburization of cast iron to produce high carbon malleable iron (< 0.5% cc) or medium carbon steel (> 0.5% cc) in the production of these specimens of armor is suggested by their distinctive lack of slag inclusions. The production of these specimens at different Romano-British forges and fineries at various dates over a period of a century and a half suggests the widespread use of the strategy of producing malleable iron and medium carbon steel by decarburizing relatively slag-free cast iron over a millennium before the widespread appearance of the "continental method" of producing steel in Europe. At this later date, large numbers of fully martinized (quenched and tempered) weapons and tools began appearing (1400). The Roman diversity in armor-forging strategies references the even more ancient variety of edge toolmaking techniques that will likely be a component of the future archaeometallurgical investigation of early ferrous artifacts.

Pleiner's (1962) survey of early knife forms is essential to the evaluation and understanding of the production of early weapons, armor, and edge tools utilizing the technology of ferrous metallurgy. Five toolmaking strategies for making edge tools (knives and swords) characterize early Iron Age, Roman, migration period, medieval, and early modern metallurgy before the appearance of the blast furnace. Pleiner's (1962) pioneering work on early metallurgy is cited by Tylecote (1987) and provides a thumbnail sketch of these early toolmaking techniques.

The simultaneous production of all the edge tool forms described by Pleiner (1962) before the beginning of the early Industrial Revolution and its mass production of cast iron also characterizes all later efforts at edge tool production. Even the lowly all-iron ax produced by isolated smithies as recently as the mid-19th century continue to make their occasional appearance in New England collections. More controversial is the ubiquitous, simultaneous presence of the all steel (German steel – continental method) and steeled (weld steel) trade axes in tool collections and archaeological sites in eastern North America. The latter form becomes more common in the 18th century, probably in response to the sudden availability of blister steel bar stock made in cementation steel-producing furnaces, the use of which became widespread in England in the very late 17th century.

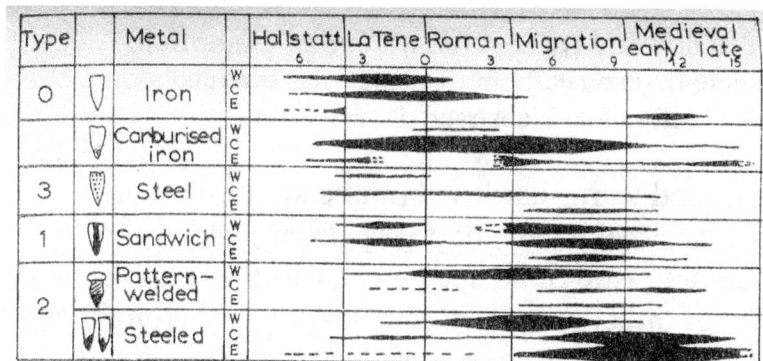

Figure 7.21 Chronology of knife forms (after Pleiner 1962).

Figure 8. Knife forms. From Tylecote (1987, 269).

The direct process for making the heterogeneous blooms of malleable iron and raw steel used for weapons and edge tools characterizes smelting practices until the appearance of cast iron in the 13th and 14th centuries. Until that date, the best edge tools and weapons, other than Damascus swords, were made from Spathic iron high in manganese, later the source of spiegeleisen (cast iron high in manganese) used to make German steel after 1350. The manganese neutralized the deleterious impact of sulfur in the direct process bloomery, facilitating its expulsion within this slag and allowing what was initially the accidental and unintended production of iron with a heterogeneous carbon content, i.e. natural steel. A high proportion of early central European iron implements circa ± 500 BC, especially those manufactured in southern Germany and Austria, was composed of raw natural low carbon steel (malleable iron with a 0.2 – 0.5% cc) rather than wrought iron. The currency bars of this heterogeneous mixture of malleable iron/raw steel were refined and reforged into steel bar stock and used for steeling the knife forms illustrated by Pleiner (1962) in Figure 8. In rare cases, insufficient carbon was present, and only an iron knife (type 0) was produced. Occasionally an all steel knife could be wrought from the reforged currency bar (type 3). Otherwise, malleable iron bar stock was carburized (type 4), steeled, or pattern-welded (type 2) by toolmaking techniques unchanged from the early Iron Age until the late 18th century appearance of rolled cast steel. The extent to which bloomery-produced cast iron was deliberately decarburized into medium carbon steel (> 0.5% cc) and/or high carbon malleable iron (< 0.5% cc), in contrast to the refining of the heterogeneous mix of the ubiquitous currency bar, is a puzzle, the solving of which will have to await much more detailed metallurgical analysis of early weapons and edge tools.

The extent of use of iron tools during the late years of Roman occupation of Britain and the turmoil that followed are illustrated not only by the wide variety of iron implements in museum collections from archaeological sites, but also the startling number of pikes, spears, and other weapons found in the Thames River and other riverine locations. Luckily for archaeologists and historians, warrior votive rituals required disposal of articles of warfare in the river. The anaerobic environment of the river mud flats served to protect these iron and steel tools from oxidation, allowing systematic retrieval of these pikes, spears, and swords by British antiquarians, who became knowledgeable about this tradition during the 19th century. The forms and styles of these implements provide a link between Roman Britain, the later florescence of Merovingian swordsmiths (± 750 AD), and the reinvigorated Iron Age that began during the period of cathedral construction (±1100 AD). That these tool forms are better documented than their metallurgy only adds to the many questions about the steelmaking strategies and toolmaking techniques used in their manufacture.

Wrought Steel: The Zelechovice Furnace

The existence of a unique early medieval community of ironmongers in eastern Europe is one of the more interesting labyrinths in our narration of steel- and toolmaking strategies and techniques before the modern era (<1870). One of the most famous writers on early ferrous metallurgy, Polish archaeometallurgist Radomir Pleiner, gives us important information on this community. Few of his writings have been translated into English, but he is often cited by multilingual American archaeometallurgist Theodore Wertime, and one important Pleiner (1969) article, *Experimental Smelting of Steel in Early Medieval Furnaces*, is available in English.

Pleiner was an avid investigator of numerous bloomery forge sites from the early Iron Age to the late medieval period, particularly those common in eastern and central Europe. Working in conjunction with the Prague Archaeological Institute and Institute of Iron Metallurgy, Pleiner investigated early medieval Slavic metallurgy, including a Zelechovice-type furnace located in northern Moravia and operated during the 8th century AD. He also correlated his experiments with investigations of the Scharmbeck low-shaft bowl furnace commonly used in northwest Germany, Poland, and Czechoslovakia during the first four centuries of the Christian era.

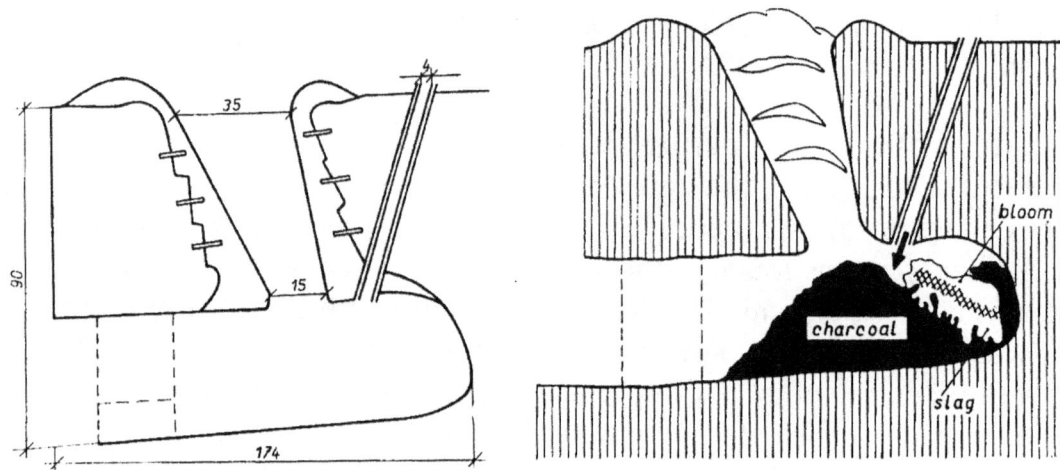

Fig. 3. Profile of the Slavic furnace of the type Żelechovice prepared for the smelt II.2. On the right hand: the position of the metallic bloom in the cavity during the process.

Figure 9. (Pleiner 1969, 461)

To perform his investigations, Pleiner rebuilt a Zelechovice furnace, which was constructed in a rectangular trench approximately 13 meters long and 2 meters wide.

32

These bloomeries often had wide slanting entrances in their midsections and nearby charcoal and ore piles and roasting hearths. Pleiner built three bloomery furnaces of this design, one of which was used for testing air circulation and the other two for the actual smelting tests. The most intriguing component of this type of furnace was the cavity in the lower furnace body, which extended in the ground to the right of the tuyère (Figure 9). The iron ore being smelted was placed in this cavity on top of and adjacent too, but not within, the charcoal fuel being fired. The most important characteristic of this particular type of bloomery furnace was its capability of producing natural steel. The bloom of iron ore being reduced was sufficiently protected from the oxidizing influences of the combustion gasses of the burning charcoal by this cavity; the resulting product of these early Moravian furnaces was often a bloom of raw natural steel with a heterogeneous carbon content.

The Pleiner text contains no discussion of whether the ironmongers of northern Moravia had access to any of the manganese-laced ores that were available to the south in the Carinthian and Styrian sections of Austria, but this type of bloomery was certainly used specifically to produce a steely iron, not low carbon wrought iron. There is, as yet, no documentation that this particular bloomery design was used in the Roman period, but the Scharmbeck type bowl furnaces, which were also used in Pleiner's investigations, are a typical form of bowl furnace used during that time. The Pleiner team was not completely successful in smelting natural steel in the furnace they built, but some steel was detected within the compact center of the bloom of smelted iron, though not in the thin sheets of sponge iron surrounding the charcoal. Pleiner sliced one of the blooms produced in the experiment into two parts using a diamond cutoff wheel, and then etched the bloom surfaces with natal. Macroscopic examination then revealed that most of the surface of the bloom was steel. Pleiner notes that there was some inhomogeneity, but it was "not quite as bad as in many original fragments of prehistoric blooms" (Pleiner 1969, 475). While all the smelts in the experiment were not successful, at least one smelt yielded a bloom that contained about 25% of the iron in the ore charge, though only part of it was of a quality that could be forged. Pleiner indicates that the necessity of taking frequent measurements of a process occurring in a relatively small furnace, in the form of partial opening of a control hole every five minutes, may have influenced the stability of the smelt. The ironmongers of Moravian prehistory would have had no such intrusions and apparently were able to produce blooms of raw steel on a regular basis. Pleiner comments that, while steel artifacts in the area he has surveyed are not ubiquitous, the objects found suggest the steel was

> ...produced in metallurgical furnaces and not by secondary cementation... [He also observes that during] excavation of the Zelechovice bloomery an unusual structural aspect of the furnace drew our attention; they were all equipped with a cavity just behind the tuyère. This arrangement was hypothetically interpreted as a reheating and

cementation chamber for the bloom, which would have been put there immediately after the smelt. The production of steel in such a type of furnace was therefore taken for granted. (Pleiner 1969, 484)

Pleiner further notes that his smelting experiments were done with "haematite" ore without manganese and with a low phosphorus content; the bloom thus produced

> …consisted of pearlite and a minimum of ferrite fibers or cementite grains: we had produced high carbon steel. Only small parts and tips were decarburized by the airflow. The hypothesis that the furnaces of Zelechovice steel thus found a splendid confirmation. (Pleiner 1969, 485)

Pleiner also comments on the many observations of carburization within medieval bloomery furnaces, as illustrated by samples of pig iron and artifacts recovered from archaeological sites. He notes that the presence of manganese was positive for carburization, whereas the presence of phosphorus had a negative influence, and also that the blooms from most furnaces had to pass through a very dangerous zone of reoxidation near the tuyère mouth. This reoxidation process reduced the carbon content of the bloom from raw steel to malleable or wrought iron. Pleiner then makes the final comment that, in contrast, in the Zelechovice type furnace

> …the reduced red-hot bloom passed relatively quickly through the oxidizing zone and slipped into the back cavity where the reheating process took place, under very good reducing conditions. The bloom, surrounded by charcoal, was protected against blowing, and the properties of the carbon steel were retained. This result… offers important evidence about the technical ingenuity of the early medieval smelters among the Moravian Slavs. (Pleiner 1969, 487)

No similar bloomery furnaces with recessed cavities have yet been documented in southern Europe, England, or the United States. The existence of this unique style of bloomery furnace again illustrates the diversity of steelmaking strategies, which characterize polymetalic societies in the centuries before the development of modern steel-producing technologies.

Merovingian and Viking Swordsmiths

It is of interest to students of the history of ferrous metallurgy that the advent of the Dark Ages failed to squelch the northern Europeans' ability to forge iron and steel tools. The memory of most smelting techniques survived the advent of the Dark Ages in Europe. Charlemagne, Carolingian King of France (768 - 816) and emperor of the Holy Roman Empire under Leo III (800 - 814), had access to steel swords and iron tools of equal and possibly superior quality to those used by his Roman predecessors. The surviving examples of exquisite pattern-welded and, in some cases, damascened swords are testament to the presence of a vigorous Frankish edge toolmaking industry in the communities of the western Danube and upper Rhine river watersheds. The finesse of Merovingian swordsmiths was only equaled by that of the well traveled Vikings until the advent of the German Renaissance and the appearance of German or Wootz steel damascened matchlock guns and swords. The florescence of Frankish culture at this time helped ensure the transition between the achievements of Roman ferrous metallurgists, the robust tradition of wrought ironworkers in the medieval period, and the development of steel-producing economies as the result of the construction of the enlarged blast furnaces of the 14th century.

The art of the Merovingian swordsmith was eclipsed only by the "Damask" patterns of Islamic Wootz steel sword blades and the famed Japanese swords, both of which are described in detail by Smith (1960). The Vikings soon followed the Franks with their own version of a steel sword that was also often pattern-welded but may have utilized more steel strips and less wrought iron that Merovingian examples. Smith (1960) suggests that the pattern-welded swords of the Merovingians were not derived from Roman prototypes, but have their own unique style based on a sophisticated pattern welding technique, which utilized strips of steel and wrought iron. Treatment with a reagent discloses the unique microstructures of Merovingian swords, which Smith (1960) illustrates in the opening chapter of his text. The later development of a heavier all forge welded steel Viking sword was probably influenced by Merovingian prototypes. Also influencing the design and forging of Viking swords was their knowledge of and contact with the sophisticated Muslim Wootz steel-producing communities to the east of the Danube, which were accessed by the Black Sea trading routes the Vikings utilized for over a century.

Also of particular interest are the other trading routes of the Vikings, one of which followed the ancient amber route from the Baltic to the Danube and through alpine passes to Venice. The Danube River itself was one of their most important trading routes, providing easy access to the Black Sea, Constantinople, and Levant. Controversy exists as to whether the Vikings forged a distinctive all steel sword that was possibly derived from their contact and trade with Wootz steel-producing Muslim cultures in the east. The extent to which their sword production techniques were influenced by Merovingian techniques originating in the lower Rhine Valley during the time of Charlemagne is also unknown.

The only evidence for a Viking settlement in North America is at L'Anse aux Meadows, Newfoundland, where excavation revealed distinct evidence of a primitive iron forge (1000 AD). However, no tools survive from this site. The Museum of London and the British Museum contain excellent examples of Viking era edge tools and weapons. Their shapes and designs reflect a long tradition of northern European metalwork differing only slightly from Celtic and Roman Iron Age prototypes. One of the distinctive forms of Viking tools is a three-in-one tool: adz, ax, and mattock, used for the convenience of raiding Vikings who needed to repair a ship, hack off the head of an enemy, or prepare a campsite using this essential all-in-one portable tool kit, in which steeled edge tools also served as weapons. The indigenous rock ores of Sweden and Norway supplied Viking swordsmiths since the beginning of Viking raiding expeditions in the late 8th century. Other sources of steel utilized by Viking blacksmiths, who were active traders throughout the Mediterranean region, remain a subject of controversy. Along with Merovingian swordsmiths, Viking ironmongers and toolmakers and their finesse at edge toolmaking, seafaring, and exploration, constitute a link between Roman and Romano-British metallurgy and the rise of early modern steel- and toolmaking strategies and techniques after 1300.

Steel Production in Europe 1300 - 1740

The shaft furnace of late medieval Europe had something in common with those used in Europe in the early Iron Age in the days of the Roman Empire: they were difficult to control. Pure wrought iron for the production of hardware, or malleable iron for horticultural tools, might have been the objective of the smelter, but, as a matter of

Figure 10. From the collection Amsterdam Archaeological Depot. Wiard Krook of the bureau Monumenten & Archeologie states this ax was "...excavated in the 1970's in the City of Amsterdam, by our Archaeological Department, during the construction of the Amsterdam subway. As the metro-trace did cross the former 15th and 16th century ship wharf's area (called the Lastage), many excavated tools could be related to these shipbuilding activities and maritime history." "material: iron; length: 14,2 cm; archaeological dating: 1575-1650; marking: Chotic A in circle; excavation date: 1973; location: Weesperstraat 55-71, Amsterdam; inventory number: MWE5-112"

This is an early example of a Dutch maker-signed edge tool, circa 1575. The maker's mark was probably placed on this ax as a way to advertise the products of the maker, either as a member of the blacksmith's guild or as an individual vendor - a tell-tale sign of an emerging market economy where toolmakers would produce and sell their tools on the open market rather than being controlled by a King, Lord, etc. This ax lacks any obvious signs of a weld steel edge, predating the widespread use of the cementation furnace to produce steel. Rather, this ax typifies the production of edge tools from the decarburization of cast iron (German steel – continental method) in the early modern period.

normal operations, most shaft furnaces produced a certain amount of liquid cast iron, as a bloom of iron absorbed carbon, and melted. In between the melting of iron with a 3 - 4% carbon content at 1200° C, which produced unwanted cast iron, and the solidifying bloom of wrought iron (with a lower carbon content 0.02 – 0.8% but with a much higher melting temperature, 1500° C) ubiquitous heterogeneous blooms of malleable iron (0.08 - 0.5% cc) would form, as well as the less common production of raw steel (0.5 - 0.8% and occasionally higher carbon content). As shaft furnace height increased, the difficulties of controlling the smelt also increased.

The development of the high-shaft Stucköfen furnace in southern Germany (1350 – 1450) was of particular importance in facilitating new strategies for producing malleable iron and natural steel. Isolated bloomeries had produced natural steel of varying quality throughout the medieval period. Individual blacksmiths transmitted their knowledge of how to make steel tools from generation to generation based on the color, texture, fracture and feel of natural tool steel (≥0.5% cc) and their growing awareness and intuitive knowledge of the importance of rapidly quenching and then tempering edge tools, armor, knives, and weapons. The evolution of the Stucköfen furnace is clearly linked with the increasing use of the strategy of decarburizing cast iron to produce steel and malleable iron.

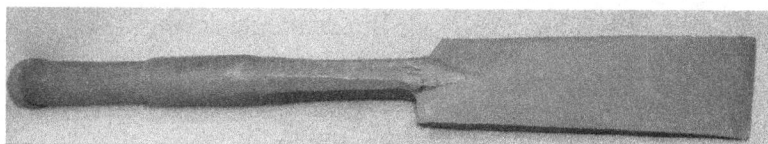

Figure 11. A large hand-forged malleable iron and natural steel slick from the 17th or 18th century. 11" long, 4 ¼" wide, in the Davistown Museum MI collection ID# 102904T1.

Producing natural raw steel was nothing new; it had been done either deliberately or accidentally since the beginning of the Iron Age. Natural steel in the form of both relatively low carbon malleable iron (0.2 - 0.5% cc) and higher carbon natural steel (≥ 0.5% cc), which could be hardened by quenching to produce martensite, then tempered to produce various pearlite-ferrite-cementite microstructures, was first produced from relatively low-shaft furnaces in the form of raw steel, which was then removed for further processing. As shaft heights increased during the late medieval period, so too would the inadvertent or deliberate production of cast iron, which would have accompanied any attempt to produce raw steel. As the high-shaft furnace grew a little taller and ran a little hotter, the manganese-laced cast iron that it produced (spiegeleisen) had to be decarburized to produce "German steel." No documentation exists to tell us when some clever ironmonger noticed that he could produce natural steel not only from the bloom of a direct process shaft furnace but also by reheating, refining, and decarburizing the salamanders of cast iron that the larger shaft furnaces were creating as an unwanted waste product. It is now increasingly evident that knowledge of how to decarburize cast iron to produce relatively slag-free, high carbon malleable iron and medium carbon steel may have been an iron- and steelmaking strategy utilized by Romano British armorers.

While some steel tools could be made using the time-consuming technique of carburizing wrought iron or that part of a wrought iron implement to be steeled, Stucköfen-produced natural steel, however impure and imperfect, could be made in relatively large quantities, e.g. 25 - 75 kg blooms. It was this natural steel that could be further refined in separate finery furnaces and then hammered into thin sheets and further carburized by case-hardening techniques. The most important steps in making quality steel tools were repeated hammering and piling, followed by repeated heating and further hammering to expel slag and other impurities, homogenizing the naturally heterogeneous iron and steel bloom into an iron alloy with evenly distributed carbon content. The appearance and development of the water-powered trip hammer (c. 1350), and the simultaneous development of the blast furnace were essential components of the rapid increase in the production of tools and weapons in the late medieval and early Renaissance eras.

Rapid quenching after heating was the final essential step in making durable steel edge tools, including weapons, knives, wood chisels, or horticultural edge tools, such as scythes and bill hooks. This was followed by tempering (reheating) and then slow cooling to relieve brittleness. Endless variations in hammering and heat treatment techniques produced wide variations in the microstructures of the steel being forged. The chemical makeup of every iron or raw steel bloom being forged differed according to carbon, slag, alloy (e.g. manganese), and contaminant (sulfur, phosphorus) content, all of which created subtle differences in microstructural formations. The widespread appearance of fully martinized steel tools also coincided with the rapid advances in ferrous metallurgy technology at this time. Blast furnace development, the use of the trip hammer, and sophisticated strategies for decarburizing cast iron to make steel constituted the technological basis for the success of the European Renaissance and the age of exploration and conquest. More than five centuries elapsed between the perfection of German steel production in Stucköfen furnaces and fineries in the German Renaissance and Bain's (1939) first comprehensive elucidation of the microstructural complexities of the heat treatment of austenized steel.

The development of the Stucköfen furnace provides a unique opportunity for a major advance in steelmaking technology after 1350. Ironmongers in the Nuremberg area of Germany, and possibly in northern Italy and Spain, had access to the superior Styrian and Carinthian iron ore from the region of what is now Austria. Its unique characteristic was that it had a manganese content of ± 2%. When smelted in the larger, high-shaft direct process Stucköfen furnace, which burned hotter but also allowed better quality control, these spathic ores neutralized the deleterious effects of sulfur in iron ore, facilitating more homogeneous carbon distribution in the smelted bloom of raw natural steel.

As the Stucköfen furnace evolved into the fully developed blast furnace, the occasional production of natural steel, which had occurred since the early Iron Age, was increasingly replaced by the deliberate production of steel from decarburized and then refined cast iron. That the cast iron produced by German smelters had a metallurgically significant manganese content was one of the key elements in facilitating the successful production of steel from decarburized refined cast iron. German metallurgists perfected this technique after 1400 or 1450, making Germany the European center of steel production during the Renaissance, until the War of the Roses decimated the German steel industry in the 4th decade of the 17th century. Moxon ([1703] 1989) has this comment on German steel, which he calls Flemish steel, illustrating its wide availability for English cutlers and toolmakers despite the presence of an already vigorous English steel industry.

> The *Flemiſh-ſteel* is made in *Germany*, in the Conntry of *Stiermark* and in the *Land of Luyck* : From thence brought to *Colen*, and is brought down the River *Rhine* to *Dort*, and other parts of *Holland* and *Flanders*, fome in *Bars* and fome in *Gads*, and is therefore by us call'd *Flemiſh-ſteel*, and fometime *Gad-ſteel*. (Moxon [1703] 1989, 58)

As the high-shaft furnace evolved into the blast furnace, the refining of decarburized cast iron resulted in the perfection of a continental "German" steel manufacturing strategy that was transferred to England at least as early as the 16th century by immigrating French ironmongers (Cleere 1985). They also brought their relatively new blast furnace technology to the forest of Dean and the Sussex Weald, where direct process bloomeries producing malleable iron and small quantities of natural steel had been operating for centuries. For two centuries, the partial decarburization of cast iron into "German steel" was England's primary steel-producing strategy. The cementation furnace came to England by 1686 (or possibly earlier). Totally pure steel was not produced until Benjamin Huntsman rediscovered the art of crucible steel production in England in 1742. Again, Moxon ([1703] 1989) has this comment on the superiority of English steel produced in the Forest of Dean where manganese-laced ores were still available in the 17th century.

> The *Engliſh-ſteel* is made in feveral places in *England*, as in *Yorkſhire*, *Glouceſterſhire*, *Suſſex*, the *Wild of Kent*, &c. But the beft is made about the *Forreſt of Dean*, it breaks Fiery, with fomewhat a courfe Grain But if it be well wrought and proves found, it makes good Edge-tools, Files and Punches. It will work well at the Forge, and take a good Heat. (Moxon [1703] 1989, 57)

The growing diversity of steelmaking strategies between the dominance of German (continental) methods before 1700 and the later evolution of England's vigorous steelmaking industry after 1700 necessitate a review of one of the most important components of the rise of modern steelmaking technology, the Scandinavian community of ironmongers who built Europe's first charcoal-fired blast furnace (Lapphyttan, Sweden ± 1250 AD).

The Importance of Scandinavian Iron

The relationship of Scandinavian metallurgists with artisans who built the cathedrals of England and France in the 11th and 12th centuries or with the early Renaissance steel production centers in western France, southern Germany, and central Europe remains unclear. Scandinavia, particularly Sweden, was an important source of high quality bar iron and possibly some direct process forged natural steel after 1400. Imported Swedish bar iron supplemented English production of wrought and malleable iron from direct process bloomeries and fined blast-furnace-derived cast iron between 1450 and 1650 and then became especially important to English steel producers once the cementation furnace was utilized to produce blister steel from wrought iron bar stock in the late 17th century. First appearing in the northeastern region of England near Newcastle, the blister steel furnace required high quality, low phosphorus, low sulfur, charcoal-fired iron to produce the best quality blister steel. Only Swedish iron would fit this need. English iron ores were high in both contaminants, making them unsuitable for edge tool production. After 1760, when the substitution of coke for charcoal as fuel occurred in England and much of western Europe, but not in Sweden with its vast forest resources, cast iron production increased due to higher blast furnace temperatures and the more efficient removal of contaminants as slag. Due to contamination with sulfur and phosphorus, the fined iron bar stock produced from cast iron using English ores was unsuitable for edge tool production. Whether for swords, knives, or woodworking and horticultural edge tools, steel made from carburized Swedish iron bar stock was superior to "German steel" made by the continental method of partially decarburizing the sulfur- and phosphorus-laced cast iron produced in charcoal, then coke-fired English blast furnaces. In contrast to Sweden, by 1700, England had no extensive deposits of rock ores uncontaminated with these two ubiquitous microcontaminants.

English forests were already so depleted that they limited the ability of English iron masters to smelt iron from charcoal. The high quality charcoal iron from Sweden became a most valued import. In contrast to forest depleted England, the vast forest resources of Scandinavia, combined with its low sulfur iron deposits, allowed Sweden, in particular, to be the most important source of high quality refined bar iron for both cutlers in Sheffield and its growing edge tool industry. The rapid increase in wooden shipbuilding in Europe after 1550 may be directly correlated with the increased availability of high quality, low sulfur, Swedish iron bar stock carburized by edge toolmakers into the woodworking tools of the shipwright and to the increasing production of "German steel" by the fining (partial decarburization) of charcoal-fired cast iron. Both sources of steel used for forging the tools of the shipwright played a key role in the exploration and settlement of the American continent and the expansion of European trading empires to all areas of the world.

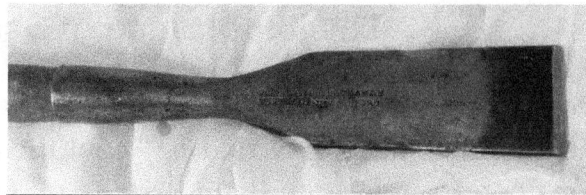

Figure 12. (top) This B. D. Hathaway slick is an excellent example of a weld steel edge tool using imported Swedish malleable iron, photo courtesy of Dave Brown. (left) Socket chisel and close up of mark made by B. D. Hathaway of New Bedford. Cast steel, 13 3/4" long, in the Davistown Museum MIII collection ID# TCC2003.

An important observation needs to be made about the type and microstructure of iron imported from Sweden before, and especially after, the widespread use of the cementation furnace in England (1685) and later in the American colonies. The iron most sought by edge toolmakers was not "wrought" iron (< 0.8% cc), always characterized by the fibrous components of its silicon slag inclusions, but the carefully refined charcoal-fired blast furnace-derived malleable iron with a sufficiently high carbon content (0.1 – 0.4%) and characteristic pearlite-ferrite microstructure that, in the modern era, would be called low carbon steel. Barraclough (1984a) provides this description of the Swedish iron bar stock so sought after by blister steel smelters.

> All the irons from the North which are sought by the manufacturers of cementation steel are distinguished by their grainy structure, compact and with a brilliant blue-grey colour somewhat resembling that of zinc. One sees most often in the transverse section of a bar all the features of a very pronounced lamellar structure and very rarely that of a fibrous structure [characteristic of low carbon silicon-laced wrought iron.] In this latter case, when nicked cold, instead of breaking with an almost flat fracture, it tears away in fibres composed of a number of superimposed plates. The surface of these plates is a slightly silvery matt white colour; their edges, when distorted after breaking cold, present a silky gleam similar to that given, under the same conditions, by the fracture of refined copper. It is extremely difficult to break the bars cold, even when heavily nicked with a steel chisel. (Barraclough 1984a, 180-1)

This Swedish malleable iron bar stock so highly valued by English steelmakers was also the same as that used by American colonists to make edge tools when they began constructing their clandestine cementation furnaces after the end of Queen Anne's War and the Treaty of Utrecht. The use of imported Swedish charcoal-fired malleable iron continued well into the 19th century. One finds numerous references in American regional and local history texts that note the importation of Swedish and, to a lesser extent,

42

Russian and Spanish bar iron to North America, especially to New England. The impost records of the New Bedford custom district, made available to this author by the New Bedford Whaling Museum, are filled with the record of multiple yearly arrivals of ships (1816-1831) laden with Swedish bar iron, which were loaded in Gottenberg and unloaded in New Bedford for use by its many whalecrafters and edge toolmakers (see Figure 12).

Once the cementation furnace became the principal source of English steel after 1690, high quality Swedish charcoal iron bar stock was the most important and perhaps sole ingredient, along with charcoal dust, used to produce high quality blister steel. After a ten day firing, the typical cementation steel furnace could produce 3 – 5 tons of blister steel, a significantly larger quantity of steel of a more uniform quality than that previously produced by the decarburization of blast-furnace-derived cast iron. When blister steel became the principal constituent of cast steel production after 1750, English steelmakers continued to use Swedish iron bar stock well into the 19th century.

The lingering perception that all such imported iron, even that from Spain or Russia, was "wrought iron," is incorrect. The higher carbon malleable iron so sought after by English and American steelmakers was the essential ingredient in successful edge tool forging. It is ironic that shipsmiths, whalecrafters, and edge toolmakers did not yet understand the essential role of the carbon content of the malleable iron used to produce the steel for their edge tools. This is not to assert that all imported iron was high carbon malleable iron. The following list of the types of Swedish iron available to English and American toolmakers (Figure 13) illustrates the diversity and invisible microstructural variety of the many grades of both wrought and malleable iron produced by Swedish (and Norwegian, Russian, and Spanish) smelting furnaces. Each and every iron type in this list must have had its unique special purpose applications for tool and sheet metal production, as well as a variable carbon content that was unknown to the ironmongers of the day.

The blacksmiths and toolmakers who utilized this broad array of iron products had only their empirical and intuitive knowledge to tell them which kind of iron was suitable for a particular toolmaking function or other use. For comments on the ability of toolmakers and swordsmiths to evaluate the quality of the hot iron they were forging see the definition of "kan of the ironmonger" in the *Handbook for Ironmongers* (Brack 2008). Blister steel produced in the cementation furnace for the specific purpose of making crucible steel and edge tools required the most refined and, thus, the highest grades of iron in this listing.

Implicit in the production of blister steel was the growing knowledge of the technique of case-hardening utilizing cementation furnaces, the size of which allowed steel production in much greater quantities and with better quality control than either steel produced by the continental method of decarburizing cast iron (German steel) or natural steel

Current prices of Swedish, Norwegian, Russian, and English irons used in the Yorkshire steelworks

Swedish and Norwegian forges

Lofsta and Carlholm (Uppsala)	£35. 0. 0. per ton
Gimo and Ranas (Uppsala)	£31. 0. 0.
Osterby (Uppsala)	£30. 0. 0.
Forssmark (Stockholm)	£28. 0. 0.
Stromsberg and Ullforss (Uppsala)	£28. 0. 0.
Gysinge (Gefleborg)	£27. 0. 0.
Wattholma (Uppsala)	£26. 0. 0.
Hargs (Stockholm)	£26. 0. 0.
Shebo and Ortala (Fahlu)	£25. 0. 0.
Oster Rusoer (Nadenaes)	£24.10. 0.
Elfkarleo (Uppsala)	£21. 0. 0.
Sorforss (West Norrland)	£21. 0. 0.
Hedaker (Westeras)	£18.10. 0.
Backaforss (Elfsborg)	£18.10. 0.
Soderforss (Uppsala)	£18. 0. 0.
Norberg (Gefleborg)	£17.10. 0. per ton
Hedwigforss (Gefleborg)	£17.10. 0.
Dadran (Fahlu)	£16.10. 0.
Rishyttan (Fahlu)	£16. 0. 0.
Catherineberg (Gefleborg)	£15.10. 0.
Thurbo and Wikmanshyttan (Fahlu)	£15.10. 0.
Awesta (Fahlu)	£15. 0. 0.
Ludwika (Fahlu)	£15. 0. 0.
Swana (Westeras)	£15. 0. 0.
Amoth (Gefleborg)	£15. 0. 0.
Strombacka and Swabenswerk (Gefleborg)	£15. 0. 0.
Tjarnes Nedre and Robertsholm (Gefleborg)	£15. 0. 0.
Hamarby (Gefleborg)	£15. 0. 0.
Storforss (Carlstadt)	£15. 0. 0.
Quarntorp (Carlstadt)	£15. 0. 0.
Friedricsberg (Carlstadt)	£14.10. 0.
Fagersta (Westeras)	£14.10. 0.
Sikforss (Orebro)	£14.10. 0.
Melderstein (Norrbotten)	£14.10. 0.
Snoa. Anderforss, Ericforss (Fahlu)	£14.10. 0.
Spjutback (Carlstadt)	£13. 0. 0.
Larsansjo (Westeras)	£13. 0. 0.

Russian forges

Nijni-Taguilsk (Perm)	£19. 0. 0.
Katav-Ivanovsk (Orenbourg)	£17.10. 0.
Jourzen-Ivanovsk (Orenbourg)	£14.10. 0.
Neviansk (Perm)	£14.10. 0.

English forges

Bagbarrow, Sparkbridge, Nibthwaite (Lancashire)	£17. 0. 0.
Lowmoor (Yorkshire)	£16. 0. 0.
Tividale (Staffordshire)	£15.10. 0.

Figure 13. Iron prices from 1836 and 1842 (Barraclough 1984a, 179-80).

produced in primitive charcoal-fired bloomery furnaces. A cementation furnace operating for a few days produced case-hardened, relatively low carbon steel, with a steely outer surface but a softer malleable iron interior. Higher quality, more homogeneous blister steel required furnace firings as long as ten or twelve days to complete carburization of the interior portions of the several tons of bar iron in a conversion furnace. Rather than fully converting wrought iron to blister steel, the case-hardening technique allowed production of tools and knives from heterogeneous blends of blister steel, which had not been fully carburized into tool steel. After further forging and heat treatment, including additional case-hardening, which could also be done in a separate furnace, tools made from this partially carburized steel were characterized by both a high carbon, hardened steel outer surface and a somewhat softer, flexible inner, lower carbon subsurface, providing both a durable outer casing and a fracture resistant interior. In the growing diversity of steelmaking techniques between 1500 and 1800, high quality Swedish malleable iron bar stock continued to play a key role in quality edge tool production. Low in sulfur, it was also the key resource for Sheffield knife cutlers, beginning at least as early as the 15th century. That it was also used for tool- and implement-making of every description other than edge tools is implicit in the wide variety of iron types available for importation from Sweden. American iron-makers soon developed a similar diversity of products with variable carbon content and microconstituents, as illustrated by Gordon (1996) in his detailed survey of the American iron industry. With the development of the cementation furnace for steel production in England after 1686, Sweden remained the main source of charcoal iron for English

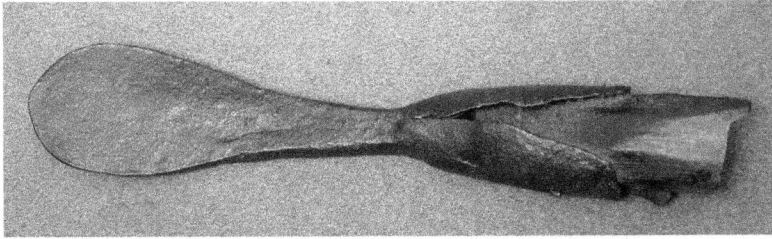

Figure 14. An interesting example of a wrought iron bark spud. The working edge was probably further carburized into malleable iron by forge welding. It is 10 ¼" long to the end of the socket, c. 1750 but typical of 17th century forms, in the Davistown Museum MI collection ID# TAB1005.

toolmakers until the late 19th century, despite the importation of significant quantities of colonial iron from Maryland, Virginia, and Pennsylvania after 1740. By the fourth decade of the 19th century, the American iron industry, characterized by an increasing quality and quantity of production, was quickly supplementing imported Swedish iron even for the raw materials needed for the highest quality U.S. domestic blister steel production.

The Accidental Origin of Cast Iron

There is a broad consensus among writers on the history of metallurgy that the blast furnace made its first appearance in Europe in the late medieval period. Tylecote (1987, 327) notes that the first carbon dated specimens of cast iron in Europe originated in Sweden between 1200 and 1300, probably produced by Europe's first documented blast furnace at Lapphyttan, Sweden. Other blast furnaces were soon constructed in the lower Rhine Valley, a center of iron production as a result of the nearby extensive rock ore deposits in the Luxembourg-German-French border region. Tylecote (1987) speculates that knowledge of the operation of the blast furnace may have been a result of the trading activity of the Swedes (Varangians) who had contact with Asian sources using Baltic, Black Sea, and Caspian River trading routes as early as the 9th century. But production of cast iron in blast furnaces was not limited to the European florescence of industrial activity in the late medieval and early Renaissance. Numerous writers, including Wertime (1980); Barraclough (1984a); and Tylecote (1987), note the widespread production of cast iron in China during the Warring States Period 475 - 221 BC and later during the Han dynasty 206 BC - 220 AD. The intriguing question arises as to whether the Chinese first produced cast iron using some variation of a furnace design specifically to produce cast iron. Needham (1958), the most noted chronicler of Chinese strategies of iron and steel production, asserts that cast iron was produced in such specially designed furnaces at least as early as 700 BC. Further consideration of the puzzle of Chinese furnace design leads to the fact that the earliest central European bloomery furnaces could consistently produce not only blooms of wrought iron but also natural steel and cast iron, depending on the fuel to ore ratio, the location of the tuyère, and the resultant temperature of the furnace during the smelt.

The occasional nodules of natural steel that appeared in a wrought iron bloom as a matter of course were one of the traditional problems bedeviling the direct process smelting of wrought iron from the point of view of the blacksmith. These nodules made working wrought and malleable iron much more difficult for the smith. Wrought iron blooms laced with these nodules of natural steel were considered an inferior product. The problem faced by the furnace master attempting to produce either high quality wrought or malleable iron or that most difficult to achieve but highly desirable product, natural steel, was that, as the furnace temperature was increased, carbon uptake by the bloom of wrought iron or natural steel increased rapidly. As the carbon content of the iron bloom increased, it passed through a series of transitional stages from being wrought iron (<0.08% cc) to malleable iron (<0.5% cc) to medium carbon natural steel (>0.5% cc) to that of being liquid cast iron with a significantly higher carbon content but a lower melting temperature than natural steel. The rapid uptake of carbon resulted in an equally rapid decrease in the melting temperature of the iron being smelted, producing liquid cast iron that ran out of the bottom of the furnace.

In the context of our understanding of the modern blast furnace, some modern commentators have commented that this liquid cast iron was always considered an unwanted waste product of the smelting process in ancient times. This techno-historic bias now requires re-evaluation. Wertime (1980) and other writers note that early direct process furnaces produced wrought iron, natural steel, and cast iron almost as a matter of course, that is, early forge masters were not able to control direct process bloomery operations in furnaces of many designs to the extent that they would only produce high quality, mostly carbon-free wrought iron. Occasional production of cast iron and natural steel seems to characterize all pre-modern iron-smelting. This may help explain why fragments of cast iron consistently appear in Roman and pre-Roman sites. Tylecote (1987) notes cauldron fragments dating to the 4th century BC from Ukraine, as well as Roman era fragments from north Wales. Tylecote (1987) speculates that these may have been imports from Asian sources reflecting the extensive trade routes of both Hellenistic and Roman times. Since cast iron was relatively easy to produce in direct process furnaces in small quantities, it is likely that ancient founders deliberately produced cast iron for the everyday utilitarian cooking pots and other uses. Remnants of such vessels would be extremely difficult to date unless located within carefully surveyed and carbon-dated archaeological sites. The assumption that cast iron vessel fragments found in Europe have Oriental origins with respect to their production sources is extremely questionable in view of the wide range of steel and iron producing strategies and techniques that now can be documented and which characterize all Iron Age societies. An example of this diversity of iron- and steel-producing strategies is the relatively slag-free specimens of Romano British armor previously noted (Fulford 2005). The use of direct process bloomery-produced cast iron, which would of necessity be subject to further refinement and decarburization, is the likely source of this slag-free steel and iron. The early roots of this more complicated process for producing steel and malleable iron, which may extend in time even earlier than Roman Britain, further complicates the simplistic division of direct process versus indirect process iron and steel production as suddenly occurring with the appearance of the first blast furnaces in Sweden and continental Europe.

The problematic nature of unraveling the multiple occurrences of deliberate cast iron production for artifact-specific uses illustrates the repeated non-sequential appearance of technological innovation by inventive human communities. The great age of Chalybean natural steel occurred at the height of the Bronze Age, circa 1900 BC. This clearly illustrates the inadequacy of the diffusion model of technological adaptation and innovation. So too does Renfrew's (1973) observation of a flourishing Bronze Age culture in the steppes of southwestern Russia (Ukraine) two millennium before the Bronze Age became well established in the Mediterranean region. In this context, the appropriate metaphor for technological change may not be diffusion in waves, or, in the case of the Industrial Revolution, diffusion by a tsunami of technological innovations, but independent eruptions of new techniques, which send out wave patterns in a non-uniform manner in both time and space. Another analogy might be the image of not quite random explosions of fireworks in a darkened landscape. At first a few explosions, then more fireworks as pyrotechnic proto-industrial society spreads across the landscape. With the coming of the blast furnace, industrialized terrestrial landscapes consist of glowing point sources (of flames, smoke, CO_2, and mercuric sulfide emissions – the onset of the age of chemical fallout) that last for months while a furnace was "in blast," a prelude to the all encompassing incandescent light of the late 19th century florescence of industrial society and of global warfare (hot or cold) that would follow after 1940.

Wertime (1980) suggests that there never was a bloomery era in the evolution of Chinese cast iron, nor do we know the exact date of the appearance in China of cast iron, which was its principal iron-smelting strategy for centuries, if not longer. Also, we may never know when the first cast iron vessels were deliberately produced in the Mediterranean region in Italy or in central Europe. It is reasonable to conclude that blast furnaces designed specifically to produce cast iron first arose as the practical response to the demand for large quantities of cast iron tools and implements and especially for cannons for the navies of the growing market economies of European empires. In China, the need for agricultural tools seems to be the most important factor. On the other hand, cast iron cooking vessels seemed to dominate the surviving examples of cast iron in pre-Roman (Hellenistic) and Roman sites. In Mongolia, cast iron cart hubs were being used as early as the 13th century (Tylecote 1987, 327). The florescence of blast furnace construction in northern Europe and southern Germany, which followed the first Swedish charcoal-fired blast furnaces, had many precedents. It should now be obvious that cast iron production for cooking vessels, wagon hubs, or agricultural tools predates the design and construction of the larger and more efficient blast furnaces that appeared in Europe after 1300. Accidental though it was, the first production of cast iron may date to the first blooms of wrought iron and is not necessarily connected to the design of the modern blast furnace. It is most important to note that the indirect process of decarburizing blast-furnace-derived cast iron to produce wrought and malleable iron and steel after 1250 AD is clearly rooted in earlier Iron Age communities, where knowledgeable smiths used the strategy of decarburizing and refining bloomery-produced cast iron to forge malleable iron and steel armor and weapons as an alternative to carburizing and forging direct process bloomery-derived wrought and malleable iron blooms, iron bar stock, or sheet iron.

The Wreck of the *Mary Rose*

Henry VIII was in a shoreline castle near Portsmouth, England, in 1545, when the newly constructed *Mary Rose*, over-burdened with troops and equipment designed to intimidate French patrols lurking off the southern English coast, suddenly capsized. The ship carpenters' tool chest that the *Mary Rose* carried to the bottom of Portsmouth Harbor when she sank is of special interest in helping to fill the gap between our knowledge of late medieval tool forms and early modern tool forms. This chest, along with the *Mary Rose* herself, has been recovered and restored at Portsmouth, England. Surviving fragments of the carpenters' tools from the *Mary Rose* are now available for study. They help illustrate the changes between medieval tool design and technology, which they reflect, and the appearance of modern-style planes in England after the great fire of London in 1666. A number of the tools recovered from the *Mary Rose*, especially including a wooden plane, which survived for over 400 years, are medieval forms similar to those illustrated by Moxon ([1703] 1989, 68) (Figure 20). Both the plane blades and the chisels from the *Mary Rose*, now lost to rust, were almost certainly forged from steel made by the "continental technique," i.e. German steel.

The *Mary Rose* fiasco marked the beginning of a renewed era of naval shipbuilding in England, which, with the help of the weather and the high quality cannons recently cast in the Sussex Weald, led to the defeat of the Spanish Armada in 1588 and ensured that the competition for North America would be a French-English affair for the next 171 years. The loss of the *Mary Rose* signified the end of a medieval era in Europe and the beginning of intensive exploration of the North American coast by English merchant adventurers and French and Dutch trappers and traders. At this time, English, Breton, Norman, Basque, and Portuguese fishermen were well acquainted with the rich cod fishing off the "newfoundlands." St. John's, Newfoundland, was already a well established late winter port of destination for European fishermen who knew that the best cod fishing was in March and April.

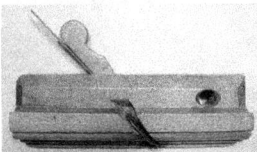

Figure 15. Robert Wooding complex molding plane, 10 14" long, 2 5/8" wide, in the Davistown Museum MII collection ID# 50402T1.

The technology of ship construction with broad ax, adz, pit saw, pod auger, drawknife, and chisel remained unchanged from the sinking of the *Mary Rose* until the appearance of the circular sawmill and steam-powered machinery (1840 - 1860) in England and America almost 300 years later. The metallurgy of edge tool production also remained unchanged but for a lesser period of time, i.e. about 200 years. The form and style of hand planes in England between 1545 and 1690 is not well documented. The great fire of London in 1664 destroyed most of the woodworking shops of the time. But, by 1690, new forms and styles of hand tools had appeared. The Robert Wooding plane (circa 1700-1710) in the Davistown Museum collection (Figure 15) signifies a new era in the production of woodworking tools. It denotes the advent of a market-based economy where individuals, such as Wooding, produced their signed tools, not for their own use or for the use of the King's shipwright but for a newly emerging market economy that was the essential stimulant for the merchant adventurers whose explorations had already resulted in the settlement of southern then northern Virginia. The merchant adventurers and explorers of the Elizabethan era and their need for wooden ships armed with cast iron cannon provided the economic stimulus for the rapid growth of English and continental shipbuilding and iron foundry industries. The hundreds of American toolmakers who would begin their task of making the tools that built America in the years after 1640 had their roots in this milieu.

The wreck of the *Mary Rose* marks a point in time where early modern iron- and steelmaking techniques (the blast furnace and production of steel by the partial decarburization of cast iron) were well established throughout Europe and England. After 1545, the rapidly expanding economies of Spain, the Netherlands, and then England and France were responsible for the further proliferation of blast furnace technology, characterized by the rise of "integrated ironworks," which included blast furnaces, finery and chafery forges, slitting and rolling mills, and blacksmith shops for making tools. John Winthrop and the Massachusetts Bay Colony established such a facility at Saugus Ironworks (Hammersmith) in 1646, providing a moderate amount of the cast, malleable, and wrought iron needed by American colonial settlers for at least a decade after this date. A more detailed review of American colonial steel and iron production strategies and options is available in *Art of the Edge Tool* (Brack 2008b). It would be another 150 years after the sinking of the *Mary Rose* before the next innovation in steel production strategies – the cementation furnace – became widely available in the late 17th century.

Steel- and Toolmaking Techniques of the Renaissance

The Blast Furnace: The First Industrial Revolution

At the same time that southern Germany was producing a natural low carbon steel with the help of its manganese-laced iron ores, larger and larger shaft furnaces were being constructed not only in southern Germany, but also in northern Europe. At some point in the 14th century, the southern German high-shaft Stucköfen furnace evolved into a true blast furnace extending the range of blast furnace operations from Sweden and the lower Rhine River valley to what would soon be the location of a great Germanic cultural and metallurgical Renaissance. Because they ran hotter and more efficiently than direct process low-shaft and bowl bloomery furnaces, large, high-shaft blast furnaces produced high carbon pig iron, not as a secondary product or as waste but as their primary product.

To produce either iron or steel from cast iron, refinery furnaces were built to decarburize the cast iron into nearly carbon-free wrought iron, low carbon malleable iron, or raw steel with a heterogeneous carbon content. One of the advantages of the blast furnace was that it more efficiently removed slag contaminants than direct process bloomeries, while lowering the high loss of iron, which occurred in the more inefficient direct process bloomery. This two step blast furnace/iron refinery process is known as the "indirect process" for producing wrought and malleable iron.

Natural and forged steel were difficult and time consuming to produce and very irregular in quality unless extensively reforged by additional hammering and heat treatments. In the Renaissance, the demand for steel for weapons and tools was greater than the supply. The Italian Renaissance in southern Europe and widespread warfare in northern Europe in the 15th century increased the demands for cannon, hand guns, steel weapons, and armor and, to a lesser extent, steel edge tools and other implements for construction, woodworking, shipbuilding, and horticultural activities. The blast furnace supplied the larger quantities of cast iron needed for the production of wrought and malleable iron and would be the later basis of English steel production that used the cementation process after 1685, which required large quantities of refined cast iron, i.e. wrought and malleable iron. As previously noted, high quality malleable (not wrought) iron produced from charcoal-fired cast iron using Sweden's low sulfur, low phosphorus ore deposits was the essential element in England's rapidly growing blister steel industry. The Swedes became world experts at fining and then refining charcoal-fired, blast-furnace-derived cast iron into high quality malleable iron suitable for the production of the best grades of English blister steel. In Germany and Austria and, to a lesser extent, in France blast-furnace-produced pig iron, especially the spiegeleisen rich in manganese, was remelted in finery furnaces and partially decarburized to produce raw steel. This raw steel was again refined, often after being broken up and again remelted and further refined into the famous German steel, which was such a superior alternative to the traditional tedious method of natural steel production or the more obscure Brescian steel production strategy. The robust German steel industry of southern Germany and Austria utilized this strategy of steel production, now known as the "continental method" (Barraclough 1984a), to meet the growing need for steel tools in the age of exploration. After 1700, its main competition was English blister steel produced by recarburizing Swedish charcoal-fired, cast-iron-derived, malleable iron. The key to increasing production of iron and steel needed by Europe's expanding market economies was the proliferation of blast furnace technology after 1400.

Brescian steel

For 200 years after the use of the blast furnace became widespread, a third steel-producing strategy, the Brescian method of making steel, was used for producing steel for edge tools, armor, and weapons, especially in southern Europe (Italy, Spain, and southern France). Bars of refined wrought iron were dipped into molten cast iron, carburizing the wrought iron, which was piled and reforged into steel bar stock, much of which was used for weapons production, especially in the city states of the Italian Renaissance. The now forgotten Brescian method of steel production, used principally in Italy and southern Europe both before and after the appearance of the blast furnace almost certainly had its roots in steelmaking traditions derived from China and Muslim communities now remembered as the source of Wootz steel.

Prior to the evolution of the modern blast furnace (1350), which operated at higher temperatures than smaller direct process bloomery furnaces and thus produced liquid cast iron as its only product, archaeometallurgical evidence indicates that cast iron was also an occasional byproduct of direct process smelting. Wertime (1980) and others clearly assert that both accidental and deliberate production of cast iron in direct process bloomeries was the frequent outcome of operating bloomeries with an increased fuel to ore ratio and with a more intense air blast. Occasional deliberate production of cast iron as a component of Brescian steel production almost certainly predated the appearance of the mature blast furnace in southern European locations, possibly as early as the heyday of ancient Noricum, the center of iron and steel production for the Roman Republic and Roman Empire. Variations of Brescian steel production strategies may have supplemented natural steel production for sword cutlers by the second century BC in southern and central Europe. Though this observation is only speculation, there may have been a close link between the mixing of wrought and cast iron to produce crucible cast Wootz steel in small kilogram quantities in ancient Indian and Muslim communities and the subsequent evolution of the Brescian strategy for steel production.

The principal use of Brescian steel in the early Renaissance was for weapons production. Bars of heterogeneous Brescian steel, probably even less homogenous in their carbon distribution than the cementation steel produced four centuries later in England, were thinned into sheets of steel, which then could be cut and welded onto or inside of iron stock to produce pattern-welded knives and edge tools. Variations of pattern welding, which combined thin sheets of steel with wrought iron, were the basis of most knife and sword production. Edge tool production by traditional techniques included the tedious forging of one tool at a time by a smith recarburizing a piece of wrought iron and case-hardening the outer surface of an iron tool, producing an implement with a hard outer surface suitable for sharpening, and a softer inner component. More commonly, most edge tools were made by inserting a piece of homogeneous steel, whether Brescian,

German, natural, or Wootz steel, into a folded piece of wrought or malleable iron and forge welding them together to make an ax or onto the edge of a socketed piece of iron to make a chisel. These were the principal toolmaking techniques that endured from the early Iron Age to the modern era.

Indirect-process-derived Brescian and German steel, more so than direct-process-derived natural steel or carburized malleable iron, were the types of steel most commonly used for making weld steel edge tools and weapons until the widespread use of the cementation furnace in England and northern Europe after 1685. A clever forge master could produce small quantities of natural steel from a bloomery furnace. Artful hammering and heat treatment by a knowledgeable blacksmith could produce a high quality natural steel edge tool, but the smith could produce only one tool at a time. The Brescian and German methods lent themselves to the widespread production of larger quantities of steel, often of equal quality to the blister steel made in England after 1685. In either case, it was the task of the blacksmith and especially the armorer to take unrefined natural steel, Brescian steel, or raw German steel with wide variations in carbon and slag content and manufacture higher quality steel by time-consuming mechanical (hammering) and heat (quenching and tempering) treatments. The widespread fame of German steel before the age of crucible steel, which was produced in England after 1750, was due to the ability of German forge masters to refine the raw steel produced from decarburized spiegeleisen and produce the sophisticated steel artifacts, as exemplified by the displays at the Victoria and Albert Museum in London, and the German firearms of the 15th and 16th century on display at the Metropolitan Museum in New York.

Written descriptions of the extent of Brescian steel production in European communities are almost nonexistent. Moxon's ([1703] 1989) description of "Venetian steel" may or may not refer to this particular technique. But the perfection of the art of decarburizing pig iron by German metallurgists during the German Renaissance to produce the famed German steel of the 15th and 16th centuries overshadowed any continental use of the Brescian strategy of steel production. In turn, the domination of blister steelmaking strategies in England after 1685 has helped obscure the importance and long duration of the art of making steel by refining and partially decarburizing pig iron in continental Europe.

Was there a more reliable way to produce a more uniform quality steel than either Brescian or German steel, especially in England where silicate and other ores laced with manganese were quickly depleted by 1700? In the 15th, 16th, and early 17th centuries, ever larger armies were battling each other throughout Europe. The Renaissance was characterized by the search for improvements in the design and efficiency of firearms ranging from cannon for the growing English, Spanish, French, and Dutch fleets of warships and merchant vessels to the need for better hand guns, the result of the increased warfare among competing nation states and nations. The European discovery of what they referred to as the "New World," the rise of competitive market and trading economies, and the rapid increase in shipbuilding, especially in and after the Elizabethan era, also played roles in increasing the demand for steel. The demand for much larger quantities of swords and armor soon expanded to include the need for tools of all kinds, such as steel-tipped edge tools for shipbuilding and the nearly forgotten horticultural edge tool, for example, the brush hooks so essential for harvesting coppice needed for charcoal-smelting furnaces. The need for wrought and malleable iron also increased rapidly after 1600. The consequential proliferation of finery furnaces that decarburized blast-furnace-derived pig iron, produced large quantities of relatively pure wrought and, especially, malleable iron the latter of which was the essential ingredient needed by the new cementation furnaces for making steel. After 1600, new centers of industrial activity based on the cementation process for producing "blister" steel first appeared both in Germany and northern Europe.

The rise of the Sheffield blister and crucible steel industry after 1700 and its important role in the evolution of the American factory system and domestic steel production has helped obscure the significance and long duration of the continental method of steel production, i.e. what we now call German steel. The same observation may be made about the now forgotten strategy of producing steel by the Brescian method.

German Steel

One of the ironies of our ethnocentric understanding of the sources and methods of steel production is our assumption that first, cementation steel, and then, after 1750, crucible steel from Sheffield, England, were the primary sources of American steel, most of which was imported, supposedly until the Civil War. The story of steel production prior to the advent of modern steel production technologies (i.e. Bessemer's pneumatic process and

Figure 16. A selection of 17th, 18th, and 19th century hewing and felling axes at the Maison de l'Outil in Troyes, France. All specimens appear to be made from German steel, i.e. steel made using the continental method of decarburizing cast iron. Welded steel cutting edges, if any, are invisible on most specimens.

Siemens' open-hearth process) is more complicated than this assumption would lead one to believe. The Sweden (malleable iron) - Newcastle - Sheffield (England) steel-producing network is well documented. The story of German steel is much less well documented and is now a nearly forgotten chapter in the history of steel production. The spathic ores of Erzberg, Austria, near the modern border of southern Germany, both of which were part of the Hapsburg Empire, had facilitated the production of natural steel tools since the early Iron Age. Halstadt, the earliest center of central European ferrous metallurgy, and Erzberg, the Iron Mountain with its siderite (manganese-containing) iron ore, were both located in the ancient Roman province of Noricum in Austria and testify to the long history of natural steel production in this area. The manganese content of the spathic iron ores of this region facilitated a more uniform carbon distribution in the smelted bloom of iron by combining selectively with sulfur at a lower temperature as slag, which was then expelled during the smelting process. The resulting bloom of iron had a more uniformly distributed carbon content facilitating natural steel production, which had been taking place in this area since the early Iron Age.

With the introduction of the blast furnace, which evolved from the high-shaft Stucköfen furnace, the use of manganese-containing iron ore from the Erzberg deposits resulted in the production of spiegeleisen, a cast iron with 5 - 10% manganese content. The partial refining of this type of cast iron produced "German steel" rather than "natural steel," both of which required extensive additional forging in order to make high quality edge tools. The cementation furnace always produced steel by the carburization (adding carbon) of malleable iron. In contrast, German steel was produced by the partial decarburizing of spiegeleisen. The decarburizing process was halted before low carbon wrought iron was produced, leaving an iron alloy with a variable carbon content, often between 0.2 – 0.8%, i.e. steel. The combination of a Celtic tradition of producing natural steel, well designed finery or smaller shaft furnaces, knowledgeable forge masters, and cast iron high in manganese (spiegeleisen) resulted in the production of large quantities of German steel. During the high Renaissance, the availability of German steel via the Rhine, Rhone, and Danube rivers must have supplanted the use of Brescian steel in

Producing Country	Natural Steel	Cementation Steel	Total Product
Great Britain	nil	20,200	20,200
Austrian Monarchy	12,600	nil	12,600
German Customs Union	7,000	100	7,100
France	3,320	3,710	7,030
Russia	520	2,640	3,160
Sweden and Norway	1,970	890	2,860
Spain	200	100	300
Italian Peninsula	100	100	200
TOTAL	25,710	27,740	53,450

300

Figure 17. (Day 1991, 300). European steelmaking in 1840, the production quantities are stated in tons. In this figure "natural steel" refers to "German steel" produced by the partial decarburization of cast iron in special finery furnaces.

the years before the widespread use of the cementation furnace in England and northern Europe became a third source of steel. German steel produced by decarburizing cast iron is frequently called "natural steel" both in old and recent literature (as discussed below). This is incorrect because "natural steel" is a direct process product produced only from the bloomery. In the south German-Austrian region, with its manganese containing iron ores, it was inevitable that the robust and ancient tradition of producing direct process

57

natural steel would evolve into the continental method of decarburizing cast iron. Though the first documented cementation furnace appeared in Nuremberg in 1601 (Barraclough 1984a), most steel produced in the southern German-Austria region in the next 250 years was produced by the continental method of decarburizing pig iron (German steel), not by the use of the cementation or conversion furnace.

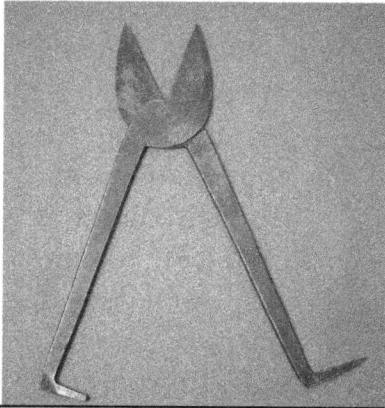

Figure 18. These forge welded malleable iron and German steel tin snips are clearly continental in origin. 15" long, 3 3/8" cutting blade, in the Davistown Museum MIII collection ID# 090508T8.

The high manganese content of the iron ores being used in the production of German steel, characteristic of the ores of the nearby Erzberg formation, played a role in both promoting homogenous carbon distribution and also in strengthening the steel being produced. Almost 500 years after the first appearance of blast-furnace-derived German steel, spiegeleisen again made an appearance in the 19th century as an essential ingredient in bulk-processed Bessemer steel. With modern knowledge of the chemistry of steelmaking, manganese is routinely added as a constituent not only of Bessemer steel but also in many types of alloy steels, which were invented during and after the era of Bessemer steel.

German steel was also a likely cargo in coasting traders from America returning from Europe in the centuries before bulk steel production. In the 1840s, the tonnage of German steel production was approximately equal to the English production of cementation steel (Figure 17). Its use remained widespread in Europe despite the appearance of the cementation process in the late 17th century. It is important to note that German steel is still misnamed "natural steel." By 1840 most "natural steel" was derived from decarburized cast iron, now more appropriately named German steel. German steel is one possible source of the many edge tools not marked cast steel that frequently appear in the tool chests of American shipwrights prior to the Civil War and the era of mass production of crucible steel edge tools. Sheffield, England, was not the only source of steel for the American shipwright.

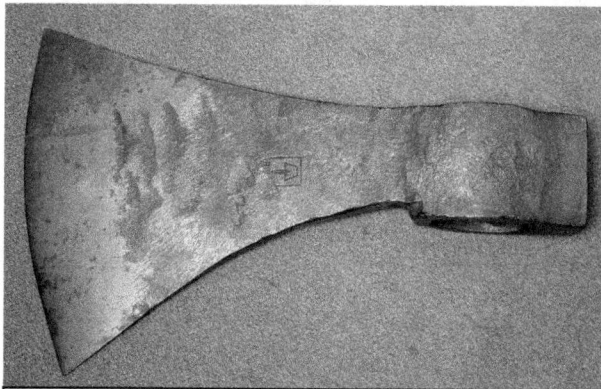

Figure 19. A malleable iron or German steel ax, which has been further carburized by subsequent forging. 11 5/8" long, 7 ½" wide, in the Davistown Museum IR collection ID# 12900T7.

The tool forms illustrated in Moxon's ([1703] 1989) *Mechanick Exercises* show the transition from late-medieval tool styles (e.g. the planes with the looped handles and the hatchet) to early modern forms (chisels, saws, planes with wedges, bevels and try squares). While we cannot be sure from the Moxon text, the edge tools are almost certainly forge welded combinations of malleable iron and German steel. The tools recovered from the wreck of the *Mary Rose* illustrate the widespread availability of German steel as a source of weld steel before the advent of blister steel production after 1650. The popular conception is that the steel in the tools of American explorers and colonists was English in origin. In forest-depleted England, coppice, rather than oak, fueled English blast furnaces after 1550. Steel was produced by the continental method until the late 17th century, when cementation furnaces began replacing the older strategy. This transition was well underway by 1686, when ironclad proof of its use is noted by Barraclough (1984a). During this transitional period Newcastle, in northeast England, became England's primary steel-producing center (1675 – 1750). After 1750, Sheffield equaled, and then surpassed Newcastle and Birmingham as the center of English steel production.

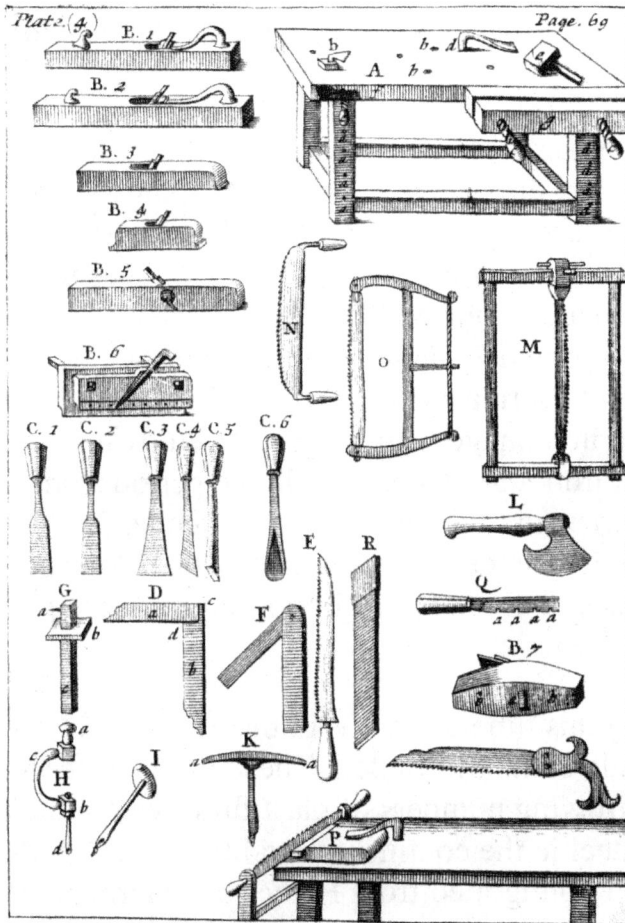

Figure 20. (Moxon [1703] 1989, 69). This illustration shows the typical tools and their designs in the 17th century. With the exception of the loop handled hand planes, most of these tools are characteristic of those brought to the colonies by the earliest English settlers. The loop handled hand planes were an obsolete medieval design already supplanted by modern Dutch-derived forms at the time of Moxon's 1703 edition.

In the period between the sailing of the *Mary Rose* in 1545 and the appearance of the cementation furnace after 1685, German steel dominated steel production despite the continued use of other steelmaking strategies. With the appearance of blister steel production after 1685, the wide variety of steels that could be produced directly from the cementation furnace as a function of firing time or indirectly produced by reforging greatly expanded the diversity of steels available for the growing number of applications for its use.

The Legacy of Multiple Steel-producing Strategies

Steel production strategies in the 18th and 19th centuries were much more diverse than what would have been the case had all steel been produced in Sheffield by a uniform series of processes, i.e. cementation - shear - crucible. The reality was that multiple strategies of steel production were being used at the same time. Some were centuries old and only produced small quantities of steel, as with natural steel production, which was still made in both Catalan bloomeries in southern Spain and American bloomeries until the late 19th century.

The extent and termination of the production of Brescian steel is nearly undocumented; even the beginnings of the use of the cementation furnace are only vaguely known. The appearance of German steel made in fining furnaces from cast iron pre-dated the appearance of the cementation furnace. The variety of steel production techniques that had evolved by the early 19th century was further complicated by Henry Cort's improvements to the refractory furnace, which provided yet another method for decarburizing cast iron to produce puddled steel. German steel produced by decarburizing blast furnace-derived spiegeleisen was greater in tonnage than that produced by the cementation furnaces that carburized wrought iron into blister steel in the period from 1650 - 1750. In fact, in the 15th and 16th centuries, direct process natural steel and German steel produced from decarburized pig iron in centers, such as Aachen, Augsburg, Nuremberg, and elsewhere in southern Germany, were the most important sources of steel in the Renaissance.

The centers of English steel production during this time period were the Forest of Dean north of the Severn River and the Sussex Weald south of London. These English steel-production areas were also characterized by growing numbers of blast furnaces producing pig iron that was partially decarburized into steel in the continental tradition, often with the help of European furnace masters who had immigrated from France and Germany (Cleere 1985). This "German" steelmaking strategy first supplanted the direct process bloomeries of the Forest of Dean and the Sussex Weald in England, and their occasional production of natural steel, two centuries prior to the advent of the cementation furnace (1685).

Smith (1968) explores the growing knowledge of the chemistry of iron and steel production up to the discovery of the role of carbon in "steeling" in 1786 by French and Dutch scientists. The literature that Smith quotes reflects the growing sophistication in steel production technologies, including advances in hardening, tempering, soldering, quenching, and the development of spring or shear steel and the cementation process. Perhaps the most startling aspect of Smith's survey is the wide variety of continental European sources for production of pure "weld" steel. Though of widely varying quality, steel, at least in small quantities, seems to have been produced everywhere in Europe after 1600. Available in small as well as large quantities, this "weld" steel from German fineries and continental cementation furnaces may have been exported to the American colonies, where colonial blacksmiths could combine direct process open-hearth-produced wrought and malleable iron bar stock with the small piece of expensive, but absolutely essential, imported "weld" steel to produce axes, adzes, drawknives, scythes, and other tools so essential for the colonization of a new frontier, the harvesting of its rich forest resources, or the construction of its ships. Blister steel from England's converting furnaces would not have been available before 1700 in the American colonies. Shortly after that date, clandestine steel furnaces, often undocumented because of their violation of English restrictions on steel production in the colonies, began appearing in American colonial ironworking communities (Bining 1933). As with American colonial era blast furnaces (e.g. Carver, MA), American colonial cementation furnaces were much smaller than the cementation furnaces utilized at Newcastle, Sheffield, and other English steel-producing centers. Most American colonial furnace operators, whether operating direct or indirect processing facilities, left no written descriptions of their furnace designs or operations. The primary evidence suggesting that small blister steel furnaces began appearing in New England and other colonies in the period between 1720 and 1760 is the ubiquitous presence of primitive domestically produced edge tools surviving from this period. That all were made with imported steel is highly unlikely.

The myth of the hegemony of blister steel imported from England cementation furnaces during the late 17th and 18th centuries to supply colonial and early American blacksmiths, ax-makers, and other toolmakers is an ethnocentric conceit. The mystery surrounding the flowering of a vigorous New England shipbuilding industry, beginning in 1645, is the extent to which knowledgeable shipsmiths and edge toolmakers might also have used a second option to make steel in small quantities, i.e. by melting and then decarburizing readily available cast iron, often derived from bog iron, in the continental tradition to make steel tools before the widespread appearance of blister steel after 1710. Savvy New Englanders in their coasting vessels knew that there were multiple international sources for the steel bar stock needed in the rapidly expanding American frontier. They also knew that a growing and now nearly invisible domestic steel-producing industry was beginning to provide competition with imported steel, especially after the end of Queen Anne's War (1713). Blister steel made in England's cementation furnaces had plenty of competition after it became available in the early 18th century.

The long tradition of wrought and malleable iron and natural and spathic steel production in continental Europe is especially highlighted by the excellent and gorgeous display of ornamental iron and steel at the Victoria and Albert Museum in London. The exhibit fills the entire front of the main hall on the second floor and includes many specimens of wrought and malleable iron from the medieval period through the Renaissance.

Particularly noteworthy are some of the elaborate ornamental iron and steel safes, locks, and architectural elements, the most outstanding of which are the malleable steel locks made in Nuremberg, Germany. The armada chest on display, probably dating from the 16th or 17th centuries, is similar in design and, in fact, a prototype for Captain Tew's pirate chest (Figure 21) currently on display at the Davistown Museum. This Victoria and Albert display provides evidence of the growing variety of iron and steel manufacturing techniques in late medieval and early Renaissance Europe. The Victoria and Albert collection does not include weapons or even edge tools; it does, however, illustrate the wide application of sophisticated ferrous metallurgy technologies in such trades as locksmithing, clockmaking, architecture, iron and steel ornamentation, and for numerous consumer products made by German blacksmiths. The Davistown Museum focuses on the history of edge tool manufacturing and features shipwrights' woodworking tools, but the broad array of iron and steel products produced in continental Europe before the availability of English blister and crucible steels foreshadows the florescence of American toolmakers in the mid-19th century and the unequaled production of the malleable cast iron planes of Chaney, Phillips, the Baileys, and the Stanley Tool Co. during the classic period of America's Industrial Revolution. The key element in the rapid growth of America's domestic steel- and toolmaking industries was not the diversity of earlier steelmaking traditions but the rapid evolution of a vigorous English blister steel industry after 1700 and the transfer of this technology and the steel it produced to the American colonies after 1720. Throughout the 18th century, as in previous centuries, steel was a costly and much sought commodity. Recycled steel tools were an important, if now forgotten, secondary source of steel.

Figure 21. Captain Thomas Tew's pirate chest. Formerly owned by the marine historian Edward Rose Snow, this 17th century chest illustrates the everyday use of relatively high carbon content malleable iron (not wrought iron) as the predecessor to low carbon steel. In the collection of the Davistown Museum.

Recycled Iron and Steel

A lack of a straightforward evolution of steelmaking techniques characterizes the diversity of steelmaking strategies from the florescence of the brief age of Chalybean steel at the height of the Bronze Age (1900 BC) to the late 19th century. Before the appearance of modern steelmaking technologies, the concurrent use of multiple methods of making malleable iron and steel was implicit in the wide variety of microstructural characteristics of the tools that survived from earlier times. The slag inclusions of natural steel tools helped to differentiate them from steel made from decarburized cast iron. The uniformity of the microstructure of hot rolled, nearly superplastic crucible steel is unmatched by any other steelmaking technique, with the possible exception of super-refined German and sheaf steel. Interspersed with these common steel types in any 18th or early 19th century tool collection or hoard are constantly reappearing specimens of recycled, reforged iron and/or steel rasps and other tools.

The continual use of recycled iron and steel is one of the more obscure labyrinths in the history of steelmaking. The common reuse of iron oxide-rich "bears," or slag deposits, of early Iron Age direct process bloomeries by later blast furnaces is a notable example of recycling. Reuse of iron and steel scrap in both early and modern furnaces is another. Steel rasps and files recycled into other tools are commonly found in New England shop lots and tool collections (Figure 22). Worn out steel rasps were occasionally reforged by the smith into edge tools, particularly before Henry Cort's reverbatory furnace and rolling mills made the mass production of cementation (weld-steel) and crucible steel tools possible in the 19th century. More commonly, in colonial New England and many other cultures and toolmaking milieus, recycled rasps and files, more so than any other tool form, were reforged into farriers' and horticultural tools. One of the tools in Figure 22 is an artfully reforged grafting iron. Occasional examples of the farriers' buttress refashioned from a recycled steel rasp have been

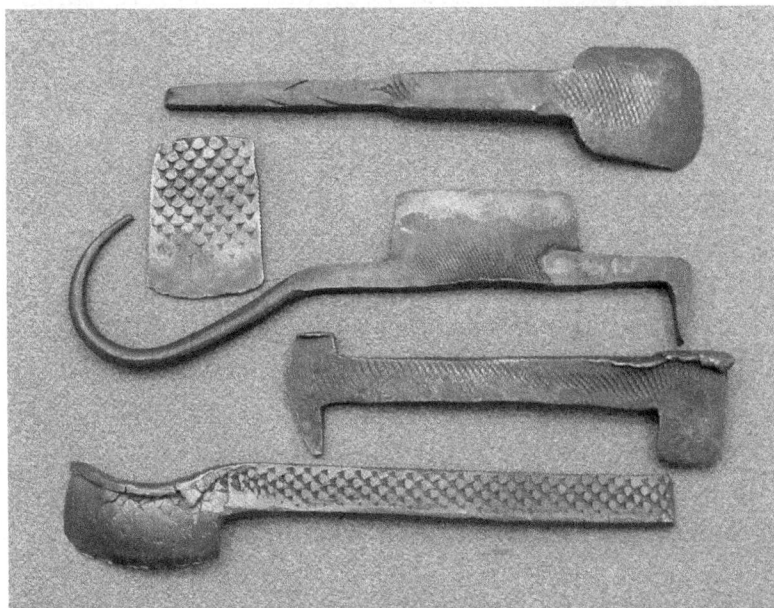

Figure 22. A group of tools made from rasps and files. From top to bottom: spud, wedge, grafting tool, and two farriers' hoof cutters. In the Davistown Museum MIII collection ID# 913108T37.

noted. Also sometimes made from recycled rasps are shoemakers' rahn (sole) files, sole knives, toe knives for hoof-trimming, shoemakers' peg cutters, and woodworkers' hook tools (for shallow face plate trimming). The Davistown Museum has two carefully refashioned drawshaves made from recycled rasps and files (Figure 23).

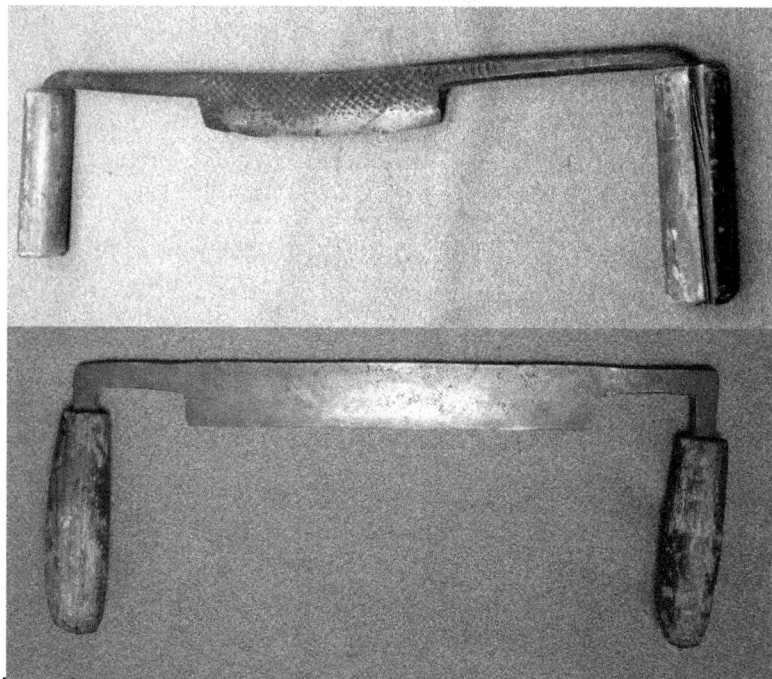

Figure 23. (top) Drawshave made from a recycled rasp. 20 ¾" long, in the Davistown Museum MII collection ID# 913108T39. (bottom) Drawshave made from a recycled file. 12" long, in the Davistown Museum MIII collection ID# 913108T23.

One of the mysteries related to recycled steel rasps and files is what the original strategy for making the tool was. Such tools were commonly made from cementation steel, as well as German and natural steel. A visual evaluation of the tool is insufficient to identify the original steelmaking strategy. At some future date, sophisticated analytical techniques may be used to differentiate cementation steel from manganese-laced German steel and silicon slag containing natural steel, but no such research has yet been published. Needless to say, in the era before bulk steel production strategies and rapid transportation by railroad or steamboat, the worn out steel rasp of the farrier was a most useful source of steel. That farriers recycled rasps into hoof-cutting, hoof-trimming, and other tools reflects the practical necessity of recycling malleable iron and steel in an era when even the thought of disposable tools, a most modern invention, would have been a diabolical conception. Though widely available as new forged steel bar stock, blister and German steels were a most expensive commodity. Worn out tools made from these steels were frequently reforged into useful, if more primitive, steel implements and edge tools. These tools were an important, if forgotten, source of steel for many a colonial toolmaker working in farm workshops.

The Cementation Furnace and Blister Steel Production

Barraclough (1984a) notes the first mention of the cementation furnace at Nuremberg in 1601. Access to the manganese-rich ores of Styria, which facilitated direct process natural steel production in small quantities from the early Iron Age to the appearance of the blast furnace, was supplemented, then replaced, after 1400 by the widespread use of the continental method of producing German steel from partially decarburized cast iron, postponing the need for alternative methods of steel production. The blast furnace greatly improved the efficiency of smelting iron. In England, it took over two centuries before the efficient production of iron from the blast furnace, sometimes decarburized into steel in finery furnaces in the tradition of the continental method, was equaled by the efficient production of steel from the cementation furnace.

The German "Stucköfen" was a high-shaft bloomery furnace producing iron by the direct process; its capacity was ± 500 kilograms of iron production per day. First appearing around 1300 AD, the Stucköfen furnace was gradually superseded by larger true blast furnaces, the flobofen (circa 1500) and the bloen (circa 1750) with capacities of 600 - 700 kilograms per day and 1500 kilograms of iron per day, respectively. The development of these larger furnaces after 1500 reflects the transition to the indirect process of iron production, in which the larger quantities of cast iron produced by blast furnaces had to be refined by partial decarburization to produce the large quantities of German steel. This was the predominant steel-producing strategy until 1700, when blister steel production in England and northern Europe began to challenge the dominance of German steel, and, to a lesser extent, Brescian steel. Southern Germany remained in the forefront of iron and steel production because their manganese-rich ore facilitated production of iron which had steel-like qualities if produced by the direct process Stucköfen furnace. With the larger blast furnaces, which produced pig iron with a 5.0 - 10.0% manganese content, the manganese became a constituent of the furnace slag, preferentially absorbing sulfur at a lower furnace temperature and, therefore, facilitating the difficult process of partially decarburizing pig iron into steel by enhancing uniform carbon distribution in the smelted iron.

In England, lack of access to the spathic ores of Styria (Austria) made natural and German steel production more difficult. The cementation process for producing blister steel was patented in England between 1613 and 1617 (Barraclough 1984) and eventually became the primary strategy for steel production in a country already dependent on high quality low sulfur Swedish iron. Much of England's iron ores were lower quality, high sulfur, high phosphorus content ores suitable for cast iron but not fine steel production. For 300 years, England relied on Swedish iron for blister steel production despite extensive use of domestic deposits for cast and wrought iron for other uses. The development of the cementation furnace allowed more control of the carburization process necessary for producing steel by recarburizing Sweden's low sulfur malleable iron produced by the fining of its charcoal-fired cast iron. Layers of fined malleable iron were interspersed with layers of carboniferous materials and the mixture was heated for periods as long as 10 to 12 days. The carboniferous materials used in the early days of the cementation process varied from charcoal dust and bone to a wide variety of strange combinations, all of which produced carbon monoxide. Essentially a closed box, the cementation furnace kept the fuel (charcoal) isolated from the ore and carburizing additives in an environment that could be checked and regulated, producing larger quantities of steel of a much more uniform quality than could be produced by the direct process as natural steel. Barraclough (1984a) translated the following quote from Jars (1774), which gives a detailed description of the design and operation of a typical 18th century blister steel furnace.

> The furnaces consist of a brick vault, about 12 ft long and 6 ft wide and 7 ft high in the middle. Some furnaces are larger or smaller. The iron fire grate on which the coal is placed is below ground under the middle of the vault. It is covered with large pieces of sandstone, resistant to fire, which form at the same time on the bottom of the chest, pot, or crucible which will contain the iron. On this base are built the sides of the crucible or chest with stone of the same kind as the base. Holes are made along the whole length of the grate which are made to come out inside the furnace between the sides of the chest and the vault. I judged that there were about six of the holes along each side so that the fire flames made at the grate were obliged to enter by these holes and envelop the whole of the chest, since the hearth and grate traversed the furnace for the whole length of the chest and flues were made each side. The flames finally debouched into the upper part in the middle of the vault where they went through a chimney flue.

> They do not put more than 4 or 5 tons of iron at most into the furnace; a continual fire for five days is needed to convert the iron into steel.

> The iron used is that from Sweden; it is known that no other is capable of making good steel. The iron is arranged in the chest with charcoal powder and the whole is covered with sand, as is the practice in Newcastle. (Barraclough 1984a, 213)

After 1700, use of the cementation furnace became widespread throughout England, and, to a lesser extent, in northern Europe, but it was used infrequently in Austria and southern Germany due to the popularity and efficiency of the continental method of producing steel by the partial decarburization of spiegeleisen. Both steel-producing strategies combined to supply a key resource for the rapidly expanding empires of Spain, the Netherlands, Portugal, England, and France in the age of exploration and settlement of the New World.

Plate 2   Derwentcote Furnace

This is the only cementation furnace remaining in the UK in its authentic form. It is stone built, with a heavily buttressed working area. It dates from about 1740 and continued in use until the late nineteenth century. Plans are now in hand for its restoration and preservation. (Photograph reproduced by kind permission of Frank Atkinson, Esq. of the Beamish Museum.)

Figure 24. (Barraclough 1984a, Plate 2. *Steelmaking before Bessemer: Blister steel, the birth of an industry.* Vol. 1. London: The Metals Society).

Sheaf Steel and the Search for Quality

Blister steel, the product of the cementation furnace, was of a higher quality and had more uniform microstructure than the natural steel produced by the direct process in smaller shaft furnace bloomeries, where pounding out a tool with steely characteristics from a bloom of iron was often a "hit or miss" situation. However, blister steel was not of a uniform quality; gaseous inclusions and other contaminants in the smelting process created blisters and air pockets in the steel. Though much of the slag associated with the production of wrought iron had been removed in the smelting of cast iron to produce the malleable iron used in the cementation furnace, the cementation process often failed to produce steel with the uniformity of carbon distribution that characterized fully martinized steel produced by the crucible process in the late 18th century. Edge toolmakers, as well as armorers and watchmakers, were aware of the wide range in the quality of blister steel, most types of which were still sufficient for many applications. However, the best woodworking edge tools, armor, watch springs, and swords required a uniformity of carbon distribution for maximum effectiveness, which most batches of blister steel could not provide.

Barraclough (1984a, 71) notes that, almost as soon as the first documented cementation furnaces were producing blister steel at Newcastle, steel sales were divided into three sub-products, i.e. "blistered or ruff steel," "faggott or drawn steel," and "gadd steel." While unrefined blister steel was the primary product of the cementation furnaces, which began operating in England at this time, the refined faggott or drawn steel production was approximately 25% of the ruff steel production. Gadd steel, which was drawn steel that had been further refined, was manufactured in quantities approximately 50% of the drawn steel production. After 1741, gadd steel was replaced by slitt steel, which also eventually replaced the drawn steel. By the mid-18th century, slitt steel production was running 5% or less of total blister steel production. The decline of gadd, then drawn, then slitt steel production between the years of 1742 and 1765 may reflect the sudden appearance of crucible cast steel, the high quality of which was even better than the highly refined drawn, gadd, and slitt steels made directly from cementation steel. The same blister steel bar stock that was used to produce these variations of refined shear steel was broken up into small pieces and inserted with charcoal dust into the clay crucibles in which crucible steel was produced.

Occasional reference to "German steel" still occurs in 19th century texts. There is an interesting footnote to English strategies for producing high quality steel. German steelmakers were brought to Shotley Bridge on the Derwent River, southwest of England's principal 17th century steel-producing center at Newcastle in 1686 and they soon utilized cementation steel produced in England from Swedish iron ores in the tedious process of folding and piling sheets of cementation steel into a more highly refined product known as "sheaf (shear) steel." The references to drawn and gadd steel in the Barraclough (1984a, 17) table almost certainly refer to the shear steel produced by the German ironmongers at Shotley Bridge. Any tool labeled "German steel" in English was produced in England by this technique, hence the misunderstanding that English sheaf steel was actually German steel. In fact, it was highly refined cementation steel made during the decades before the appearance of Benjamin Huntsman's famed crucible steel. Due to its probable lower cost, it was still produced well into the early 19th century, as indicated by the numerous English tools marked "German steel."

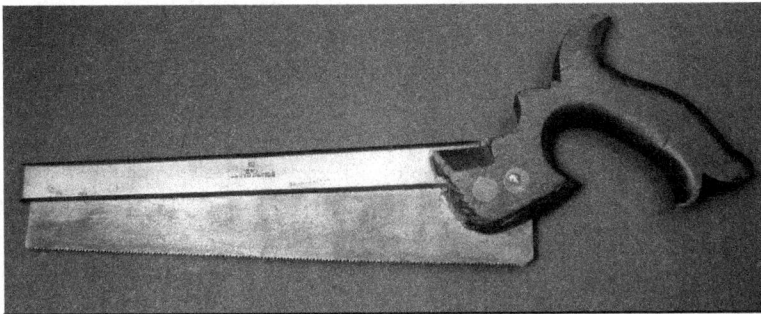

Figure 25. A backsaw made by Spear and Jackson of Sheffield, England, which is also marked "GERMAN STEEL". 23 ¼" long, in the Davistown Museum MII collection ID# TBJ3000.

The German steelmakers who migrated to the Newcastle/River Derwent area of England in the late 17th century were grounded in centuries of careful refining of decarburized cast iron into high quality German steels. Their expertise may have been the key to refining blister steel into high quality shear steel, also known as "sheaf," "spring," and "double sheaf" steel. In 1693, William Bertram, one of the newly arrived German steelmakers, was shipwrecked on the east coast of England and settled near Newcastle, where he later operated steelmaking facilities there and at Blackhall Mill, also in the Derwent valley. Initially, these facilities utilized the continental method of steel production due to small deposits of manganese-rich brown hematite in the Derwent valley. Barraclough (1984a) provides this account of Bertram's role:

> William Bertram is quoted as having pioneered the production of 'German steel' by forging blister steel. He produced five kinds, the hardest being 'Double Spur and Double Star'. Progressively getting softer, the other grades were 'Double Spur and Single Star', 'Double Spur', 'Double Shear', and 'Single Shear'. Angerstein is quite definite that the 'Shear Steel' mark – a stamp showing crossed shear blades – was Bertram's own mark; thus it comes as no surprise to learn that the making of shear steel was introduced into Sheffield by a workman from Blackhall Mill in 1767. The process was developed by building up a knowledge of how to segregate the blister steel into batches of similar hardness, presumably by examination of the fracture, using selected grades of iron. Suitable bars would then be faggotted and forge-welded. Bertram had built up a reputation for quality in this way; since he was a German, it seems to have been accepted that he had produced the true German steel –

this presumably is where the later confusion between German steel and shear steel arose. (Barraclough 1984a, 65-66)

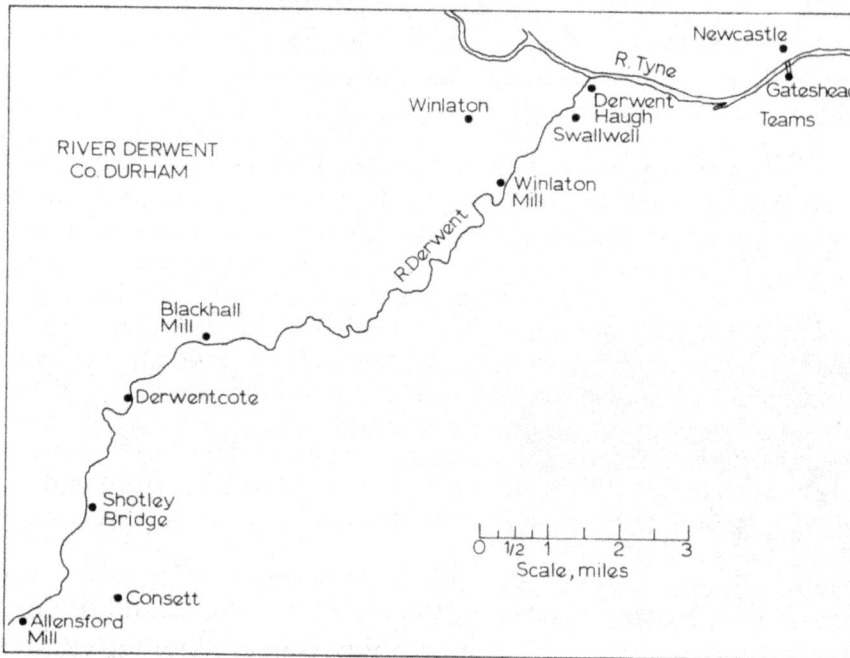

Fig. 10   Early steelmaking sites on the River Derwent

The establishment of cementation furnaces in the early eighteenth century at Winlaton Mill, Swalwell, Newcastle, Blackhall Mill, Derwentcote, and Teams made this area the major centre of steelmaking in the country. Shotley Bridge was the home of the Hollow Sword Company; it is possible that steel was produced here by the old German method in the last years of the seventeenth century. The use of pig iron from Allensford, nearby, for this purpose could be postulated. The later iron and steel centre of Consett is in this area.

Figure 26. Barraclough, K.C. (1984a, 63). *Steelmaking before Bessemer: Blister steel, the birth of an industry.* Vol. 1. London: The Metals Society.

That confusion continues today, facilitated by the mark (in English only) "German steel" on many English backsaws (Figure 25). As noted by Barraclough (1984a), this high quality saw steel was shear steel made from reforged blister steel, and its production was totally different from that of German steel. Blister steel production was tedious, time-consuming, and expensive, and it accelerated the destruction of European forests. Sheaf and spring steel production, i.e. reprocessed cementation or blister steel, represented a fourth stage in the indirect process of manufacturing edge tools from iron ore. The labor-intensive nature of piling, folding, and reforging strips of steel for special applications, such as razors, knives, saw blades, watch springs, pins, and woodworking edge tools doubled the cost of blister steel but only partially solved the need for absolutely pure steel for these products. It was the combination of the complexity and expense of production and lack of uniformity of sheaf steel that prompted Benjamin Huntsman to search for a simpler method to produce pure steel in small quantities.

Early Modern Metallurgy

Crucible Steel: The Second Industrial Revolution: Part I

The most important element in the evolution of Europe and America's growing market economies was the need for ever more precise measuring tools and, eventually, the machinery of the Industrial Revolution. This increasing demand for steel with a more uniform quality, which results from the uniform distribution of carbon within the microstructure of steel, led to the development of crucible steel. A key element in the evolution of crucible steel production was Abraham Darby's discovery of how to efficiently extract coke from coal allowing higher furnace temperatures, which glass blowers in England were already utilizing to make improved navigational aids and which would be so useful for cast steel production. English inventor and watchmaker Benjamin Huntsman had an urgent need for higher quality steel for his watch springs. Though they contained fewer slag inclusions than natural steel, blister and shear steels still had tiny particles of slag that, as constituents of watch springs, limited the possibilities for manufacturing ever more accurate and smaller watches. Huntsman may or may not have been aware of the tradition of crucible steel production in China and India or its presence as a Viking trade item or in Viking swords, a development only recently recognized by historians of metallurgy (Tylecote 1986, Wayman 2000). However, Huntsman was almost certainly familiar with the reputation of crucible steel, which was occasionally described in writings on metallurgy, the most noteworthy commentary being that of Moxon ([1703] 1989). Excerpts of Moxon's commentary on steelmaking are contained in Appendix E of the *Hand Tools in History* series volume 7, *Art of the Edge Tool* (Brack 2008a). In any case, Huntsman single-handedly reintroduced crucible steelmaking techniques in England in 1742.

Breaking up small quantities of blister steel, adding flux, and using crucibles made of special heat resistant Stourbridge clay, the same clay used by John Dollard to manufacture his achromatic lens (1757) for more accurate star sighting during navigation, Huntsman reheated and remelted the steel in ovens with no fuel contact. After a period of hours, rather than days, most remaining slag impurities were skimmed off the surface of the melted steel, resulting in the production of steel with homogenized carbon distribution. This resulted in a

Figure 27. Drawknife made by I. Pope and marked "CAST STEEL". This tool is typical of a domestically forged tool utilizing imported cast steel. 15 ½" long, in the Davistown Museum MIII collection ID# 913108T51.

uniformity of grain size throughout the steel. The microstructure of crucible steel is described as "martinized" by modern metallurgists, and its uniform microstructure marked a significant advance over both cementation-furnace-produced blister steel and bundled and reforged shear steel. The process avoided patchy and irregular microstructures resulting from the more heterogeneous distribution of carbon in the blister and shear steel. With its uniform carbon distribution, the high quality of crucible steel remains unsurpassed by any modern alloy steel for edge tool production. It was particularly useful for toolmaking because it was easily shaped by hot-rolling.

Huntsman's cast steel was also the critical element in allowing improved metal-cutting tools used on the newly invented screw-cutting lathe to shape the machinery of the Industrial Revolution before the introduction of alloy steel. It also facilitated improvements in navigation by allowing the production of higher quality steel for chronometer springs. It was also used in the production of the newly invented circular dividing engine, which allowed production of the highly accurate telescopic sight for the sextant, which required scribing 2,160 marks to complete the circumference of its measuring component (Burke 2007).

Crucible cast steel plays a critical role in the second stage of the Industrial Revolution. Initially kept a secret by Huntsman, crucible steel production for edge tools, which did not reach significant quantities before 1785, constitutes a recognizable landmark in the modern era due to the ubiquitous marking of almost all edge tools produced with crucible steel, with the notation "cast steel" or "warranted cast steel." For a century after 1785, this insignia advertised cast steel edge tools as superior to the weld steel edge tools made of steel produced by the older technologies.

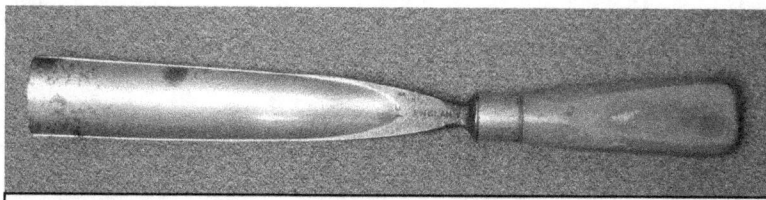

Figure 28. Patternmakers' gouge marked "S.J. ADDIS CAST STEEL" and "ENGLAND", 8" long, in the Davistown Museum MIV collection ID# TCT1002.

Exquisite English-made chisels, gouges, carving tools, shaves, knives, and other edge tools are the most obvious legacy of Huntsman's reinvention of crucible steel production. His need for high quality watch springs played a major role in providing the cast steel hand tools that built the ships, the wood patterns for casting machinery, and much of the wooden infrastructure of the coming Industrial Revolution. But crucible steel provided only a tiny percentage of the growing need for steel and iron in the early 19th century. Three other industrial developments occurred in the late 18th century that were essential components of the second Industrial Revolution. The use of coke instead of charcoal as blast furnace fuel, the development of the steam engine, and the redesigned reverbatory furnace all combined with crucible steel production to provide the basis of a massive expansion of

72

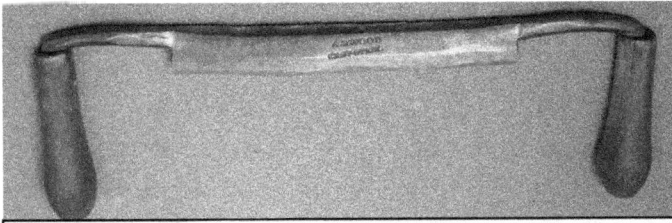

Figure 29. Drawknife marked "A.G.Wood" and "CAST-STEEL". This is an example of an American-made drawknife, probably utilizing imported English cast steel. 16 ½" long, in the Davistown Museum MIV collection ID# 071704T5.

industrial production, first in England and Europe and then, after 1830, in America. Between 1785 and 1860, English-made crucible steel, though produced in small quantities, dominated the small but essential market for high quality steel edge tools and machine components. The continuing use of and improvements in the older steelmaking strategies and technologies, especially in the decades before America perfected the art of crucible steel production in the early 1860s, remains a nearly undocumented chapter of American industrial history.

The Steam Engine: The Second Industrial Revolution: Part II

Other than the use of coke instead of wood to provide energy, the most important event in the second stage of the Industrial Revolution was the invention of the steam engine. It was the prime mover in the factory system of mass production used first in the textile industry. The steam engine was also the key component in the rapid expansion of cast iron and steel production. It was made of cast iron and first utilized effectively to pump water from mines. It was soon applied to manufacturing applications where water, especially in England, was no longer a source of energy. However, the steam engine was initially invented, not because of the lack of water-power in England, but because of the lack of wood. The shortage of wood as a heating fuel resulted, in part, because of the widespread use of charcoal for fueling blast furnaces. Coal and its daughter product coke were widely available as substitutes for blast furnace charcoal and as a home heating fuel, but, unfortunately, coal was often located in mines below the water table. The need for coal necessitated deeper and deeper coal mines, which, in turn, required the invention of pumping equipment. Thomas Savery's water-raising pump was soon followed by Newcomb's steam pump. Initially designed to pump water from coal mines, these two proto-steam engines provided James Watt with the basic principles needed to produce the first fully functional steam engine in 1763.

The steam engine is the symbol of the modern age, the mother of all machines in a rhetorical, if not in a practical, sense. It is a machine which uses water vapor as a power source to convert heat to do mechanical work. Along with the transformation of water-power into work, the invention of a tool to transfer heat into work are the two keystones of the Industrial Revolution. The mass production of cast iron by the blast furnace, which was the first essential stage of the Industrial Revolution, provided the raw materials with which to build the machines of the Industrial Revolution. The steam engine (in lieu of water-power, which was in short supply in England, but not in America) provided the link between an energy source (coal, as coke) and the thermodynamic application of this energy (heat) to the rhythmic changing of the atmospheric pressure of a volume of air, the key to the operation of the steam engine.

The steam engine could not have wide application until a substitute for charcoal fuel for blast furnace operations was developed. Coal could not be used in blast furnaces because it so thoroughly contaminated the iron being produced with sulfur. In 1709, Abraham Darby discovered how to use coke instead of coal in a blast furnace, but the large scale production of coke to fuel steam engines did not occur until later in the 18th century. Starting with Huntsman's reintroduction of crucible steel and ending with Cort's redesign of the reverbatory furnace and his invention of the rolling mill in 1784, the period from 1742 - 1784 represents the calm before the storm of an English Industrial Revolution that resulted in a cascade of new inventions or the practical implementation of recent innovations. In England, the factory system of mass production began emerging once coke was produced in sufficient quantities to power steam engines, especially those used in the textile industries.

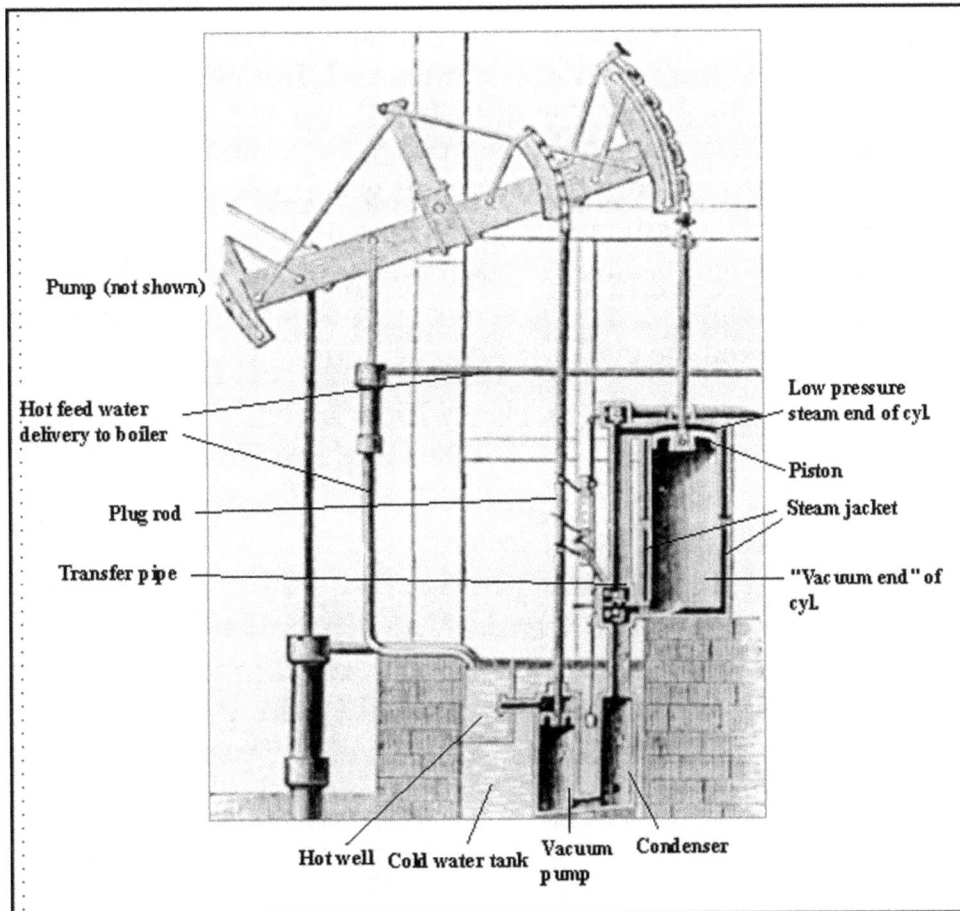

Figure 30. Watt steam pumping engine (Thurston 1878).

The roots of the steam engine can be traced back to Hero of Alexander's Pneumatica (circa 130 BC) a primitive steam reaction turbine, which was invented to open and close temple doors. After a gap of 17 centuries, Renaissance Treatises on Pneumatics began appearing (Della Porta, 1601, Giovanni Branca, 1629). In England the first prototypical steam engines were designed by the Marquis of Worcester, 1663, and Thomas Savery, who obtained a patent for a water raising engine. This engine was initially used in Britain for pumping mines, operating water wheels and supplying public water systems. Thomas Newcomb (1705) separated the boiler from the cylinder, designing a piston engine with potential for wide practical applications. In 1763, James Watt, while repairing a Newcomb atmospheric engine, noticed how the alternative cooling and heating of the piston cylinder wasted heat. Watt added a cylinder to hold the hot steam, an air pump, and an insulating steam jacket around the cylinder. Watt patented his improvements in 1769. A most important, and now almost forgotten, innovation of the second Industrial Revolution was John Wilkinson's invention of a boring machine to make the engine cylinders, which were the key component in Watt's newly designed steam engine (Figure 30). Other improvements followed - the double action steam and vacuum applications by

Watt in 1782 and the introduction of a noncondensing high pressure steam engine by Richard Trevithick in England and Oliver Evans in America in 1800. The later improvements paved the way for the development of steam-driven carriages and locomotives. Compound engines followed, and the first steam boat appeared in England in 1802. C. A. Parsons' 1884 invention of the compound steam turbine coincided with the third stage of the Industrial Revolution and its bulk steel production technology. Thus, there is a constant synergistic relationship between improvements in steam engine design and efficiency and the historical growth of iron and steel production in Europe and then in America. The combination of numerous domestic sources of water-power and the portable steam engine, the latter of which facilitated factory construction anywhere with railroad access, gave rise to America's various indigenous toolmaking industries in the second quarter of the 19th century. A century later, the sudden rise of far eastern steel-producing industrial communities marked the end of the classic period of American toolmaking.

The Reverbatory Furnace: The Second Industrial Revolution: Part III

In 1784, Henry Cort, who had just invented and patented grooved rolling mills for producing bar stock and iron rod, redesigned the reverbatory puddling furnace, which had been in use in England for almost two centuries. This new version of an ancient furnace design served the purpose of efficiently reprocessing the much larger quantities of cast iron smelted with the help of the steam engine. The reverbatory furnace thus facilitated a concomitant growth in the production of high quality wrought and malleable iron, using coke, rather than charcoal, as a fuel. Older furnace designs were plagued with the problem of fuel-ore contact. The innovative design of the reverbatory furnace separated the cast iron being decarburized from the fuel, preventing contamination of the iron in the furnace with sulfur in the fuel. The oxidation of iron being smelted due to contact with combustion gasses and inefficient contaminant oxidation had been ongoing problems in traditional refinery furnaces. In the reverbatory furnace:

> The hearth was lined with iron ore mixed with roasted puddling cinder from a previous operation; such a combination was obviously rich in iron oxide and relatively free from salacious matter. Onto this hearth was charged some 300 to 500 pounds of pig iron, which was melted down under the action of heat from a coal fire on the adjoining fire gate. (Barraclough 1984b, 92)

Wrought iron production in the reverbatory furnace went through several stages, as follows: oxidation of silicon, manganese, and phosphorus and their fixation in slag; followed by rabbling (mixing with a rabbling tool). The seething mass of liquid iron would lose its carbon content as carbon monoxide was released and burned. Decarburization of the pig iron in the furnace would eventually lead the liquid iron to become a spongy mass of wrought or malleable iron within the liquid slag as the melting point of the iron rose in proportion to the fall of its carbon content. The resultant rabbled balls of iron ($\pm$ 35 kg) would be removed from the furnace and hammered to eliminate remaining slag. The bloom of wrought or malleable iron would be reheated before rolling, shingling, or other mechanical treatment. Of particular importance for edge toolmaking was the fact that the wrought and malleable iron produced by this indirect process had a significantly lower silicon slag content than direct process bloomery-produced wrought iron and thus needed less fining and forging.

The reverbatory furnace was revolutionary in its impact, allowing a wide expansion in industrial production. It increased the quality of iron available for blacksmiths who made tools and implements prior to the era of bulk-processed carbon steel. It increased the volume of iron available for manufacturing purposes. The most important product of the reverbatory furnace was puddled iron bar stock, which was essential for the growing needs of a proto-industrial society, for iron and steel implements of every description, and for making blister steel in cementation furnaces. The need for steel during and after the Napoleonic wars increased rapidly. Coal-fired reverbatory furnaces provided wrought and malleable iron bar stock for every practical application where cast iron could not be used. One possible exception was its use for edge tool production. In Sheffield, the highest quality edge tools were produced from cementation steel made with charcoal-fired Swedish malleable iron, but this represented only a tiny fraction of the burgeoning demand for other types of steel. The steel price list of the Sanderson Brothers (Figure 31), an English company with offices in New York, illustrates the wide variety of steel that could be imported from Sheffield. In 1863, competition from domestic steelmakers, such as those in Pittsburgh, was just beginning but would surpass English production within two decades. The 26 steel types listed by the English firm Sanderson Brothers in Figure 31 are particularly significant in that they reflect the wide variety of steelmaking strategies, which were the result of centuries, not decades, of the empirical experience of ironmongers from many countries who smelted, fined, forged, cast, and reprocessed iron and steel.

Barraclough (1984b, 93) makes an important observation about Cort's reverbatory furnace: "Cort originally expected he would be able to produce steel by means of this furnace." While it took another 50 years to perfect the production of puddled steel by the decarburization of pig iron, and this mostly in Germany, enterprising English forge masters also certainly produced significant quantities of steely cast iron to construct the machinery

Table 1.2 *Sheffield and Pittsburgh crucible steel prices 1863*

	Cents per lb		
	Sanderson Bros New York	Singer, Nimick Sheffield Works	Jones, Boyd Pittsburgh Works
Best cast steel	22	21	20–21
Extra cast steel	23		
Round Machinery	14	13–15	13–16
Swage cast steel	25		
Best double shear steel	22		
Best single shear steel	19		
Blister first quality	17½		
Blister second quality	15½	} 8–12	
Blister third quality	12½		
German steel best	15½		
German steel Eagle	12½	} 9–12	} 9–11
German steel third quality	11½		
Sheet cast steel 1st quality	22		
Sheet cast steel 2nd quality	18	} 15–21	} 15–23
Sheet cast steel 3rd quality	16		
Shovel steel best	14½		
Shovel steel common	13½		
Sheet cast steel for hoes	14½	11½	11½
Mill saw steel	15½	14	14
Billet web steel	17½		
Cross-cut saw steel	17½	18	18
Best cast steel for circulars to 46 in.	25	23	23
Toe corking best	10	9¾	9¾
Spring steel best	11		
Spring steel 2nd quality	10	} 9–10¾	} 9–10¾
Spring steel 3rd quality	8¼		

Source: SCL Marsh Bros. 249/24, 28–9.

Figure 31. (Tweedale 1983, 24) Table 1.2 from *Sheffield steel and America: Aspects of the Atlantic migration of special steelmaking technology, 1850-1930*. London: F. Cass & Co.

invented by the English industrial revolutionaries who followed Henry Cort. This steely cast iron, which was perhaps puddled or German steel, can be seen today in the many early 19th century machines on exhibition at the Victoria and Albert Museum in London. German, English, and American puddled steel later played a critical role in supplying steel needs between 1835, when the steel puddling process was perfected, and 1870, when bulk steel production by the Bessemer pneumatic and Siemens open-hearth process came to dominate steel production. Puddled steel may also have played an important, but as yet undocumented, role in those critical years of the Industrial Revolution between 1800 and 1835 when steel was in short supply. Puddled steel from Cort's reverbatory furnaces may have eventually joined sheaf steel and German steel in supplementing expensive crucible steel for edge tool production from 1785 to 1865, when American-made crucible steel became widely available. One can surmise due to the long established finesse of German forge masters at producing steel by fining cast iron that individual furnace masters were able to produce steel in puddling furnaces sometime after 1785 but before the widespread appearance of bulk process steel after 1875. No written records exist to prove this supposition.

Charcoal Iron and Edge Tools

Reverbatory furnaces were fueled by coal or coke, most often by coke. Edge toolmakers had always known that iron made from refined pig iron produced in coke-fired blast furnaces was very inferior to high grade, Swedish, charcoal-fired iron used for the production of the cementation and sheaf steel used to make edge tools. The reverbatory furnace partially solved this problem by isolating the sulfur-producing fuel from the iron being puddled, thereby increasing the quality of wrought iron, which then could be used to make steel for many other uses. Nonetheless, the Swedish charcoal iron traditionally used in the first stage of crucible steel production via the cementation process maintained its reputation for excellence in the specialized production of crucible steel for gouges, chisels, drawknives, carving tools, and other edge tools for woodworking. Even after reverbatory-furnace-derived puddled iron dominated the market, most edge tool manufacturers wanted cementation steel made from Swedish, charcoal-fired, malleable iron for their crucibles. Unlike English blast furnaces, Swedish blast furnaces were charcoal-fired, as were the puddling furnaces and fineries that decarburized this cast iron and fined it into the relatively high carbon >0.08 cc malleable iron so valued by American and English toolmakers. Tweedale (1987) notes the continuing robust market for Swedish charcoal iron for crucible steel production long after the introduction of Cort's reverbatory furnace.

Figure 32. The Underhill clan were some of New England's most prolific toolmakers from the late 18th century to the late 19th century. This carefully forge welded "cast steel" and iron gouge (c. 1850) almost certainly utilized imported Swedish malleable iron, as well as imported English cast steel, in its fabrication. 13 7/8" long, 3 1/8" wide, in the Davistown Museum MIV collection ID# 112303T2.

In America, Swedish iron was a major cargo on colonial merchant ships returning to Boston and Salem. Since the cementation furnace had made its appearance in the American colonies by 1713 (Bining 1933), it is likely that indigenous, if clandestine, colonial steel production was already well underway well before the American Revolution. Early U. S. Custom's impost records for New Bedford show continuing shipments of Swedish iron to New Bedford from Gottenberg, Sweden, between 1816 and 1831. Individual steelmakers in Sheffield continued to use high quality Swedish charcoal-fired bar iron and, after 1854, charcoal-fired pig iron for crucible steel production well into the early 20th century. Only charcoal iron was suitable for production of the highest quality steel for specialized functions, such as making precision

80

measuring tools, razors, and edge tools. Crucible steel made from Swedish low-sulfur charcoal iron was the most expensive of all steels. It was also the highest quality steel available before the era of R. K. Mushet alloy steels.

An important observation needs to be made as to why charcoal-fired iron, including the famed Swedish malleable irons, played such an important role in edge tool production. The large quantity of silicon in bloomery-wrought iron (2 – 3%) is detrimental to the forging of high quality edge tools and was traditionally greatly reduced by the water-powered trip hammer or the hammer of the blacksmith, the object of the hammering being the expulsion of silicon slag. Nonetheless, charcoal iron always contains some traces of silicon, which play an important but obscure role in the production of edge tools made with crucible steel. Tiny amounts of silicon remained when the relatively pure, Swedish-made, charcoal-fired, wrought iron was made into cementation steel. The traces of silicon still remaining in cementation steel were enhanced by additional traces of silicon shed by the clay crucibles during crucible steel production. These silicon traces apparently played a key role in the metallurgy of crucible-steel-derived edge tool production for woodworkers. There appears to be a connection between silicon traces in crucible steel and the use of charcoal-fired rather than coke-fired wrought iron to produce cementation steel. Edge toolmakers traditionally used charcoal-fired malleable iron to make special batches of cementation steel for either double-refined shear steel or crucible cast steel for their edge tools. Malleability, ductility, tensile strength, and lack of brittleness are characteristics associated with crucible steel for edge tools made from high quality charcoal iron. Traces of silicon in steel made specifically for edge tools (other than knives) may have played a central role in enhancing the quality of the best edge tools.

Figure 33. A typical, finely crafted, imported, Sheffield-made drawknife, marked Spear and Jackson cast steel, 17 ¾" wide, in the Davistown Museum MIII collection ID# 112704T4.

Most of the iron fined in a reverbatory furnace was coke-fired pig iron. Documentation is lacking as to the extent to which crucible steel producers specifically made edge tools with iron that was originally smelted with charcoal, but the literature (Tylecote 1987, Barraclough 1984a, Tweedale 1986b) indicates so many subgenres of crucible steel categories that cementation steel destined for edge tool production in crucibles may have been specifically made out of charcoaled iron. When crucible steel production for edge tool manufacturing ended in the 1920s and 1930s, the era of the availability of high quality edge tools, i.e. chisels, slicks, adzes, carving tools, etc., also ended. The end of the crucible steel era also coincided with the end of the era of wrought iron production. New modern bulk steelmaking processes were responsible for the demise of the age of wooden shipbuilding, which had lingered into the first two decades of the 20th century. When the internal combustion engine replaced the unwieldy steam engine on large ships, it soon became practical to use it on smaller fishing vessels. By the beginning of the 20th century, the wooden age, already in its twilight since before the Civil War, unequivocally ended. The irony of its demise is now increasingly evident as we seek renewable energy alternatives, such as wind, to replace carbon dioxide-emitting power sources, such as the internal combustion engine.

The English Industrial Revolutionaries

Thomas Savery, Thomas Newcomb, Benjamin Darby, Benjamin Huntsman, John Wilkinson, James Watt, Richard Handbury, and Henry Cort were the first generation of inventive English engineers, the industrial revolutionaries who made possible the radical cultural and industrial changes of the 19th century. With the exception of Huntsman, who implemented a key innovation in ferrous metallurgy, these men designed or made improvements related to the two fundamental instruments of the second Industrial Revolution: the steam engine and the reverbatory furnace. The existence and efficient functioning of these two key tools of the Industrial Revolution opened the door to the design and production of function-specific machines that insured the success of the Industrial Revolution as a symphony of engines of social and economic change. Only the invention of the circular dividing engine rivals the importance of the steam engine and the reverbatory furnace. Perfected by the English engineer Jesse Ramsden, who also improved the newly invented screw-cutting lathe, the dividing engine was used for precision scaling of newly improved navigation equipment such as the sextant. Existentially, the Industrial Revolution was not just the theoretical presentation of a crystal palace exhibition of innovative equipment, but was also in its essence, the practical application of these inventions.

While Newcomb, Watt, and Cort were inventing the big ticket items of the Industrial Revolution, numerous other industrial revolutionaries were designing and inventing function-specific machines that eventually, though not always obviously, brought a gradual end to the tradition of handmade hand tools. John Kay was one of the early inventors who started the revolution in the textile industry in England with the invention of a flying shuttle (1738) that doubled loom production. James Hargrave followed with his spinning jenny for weft spinning (1764). With the help of John Kay, Richard Arkwright designed his spinning frame, which extended the functions of Hargrave's jenny, which would only weave weft, to include the weaving of the warp (1769). Arkwright opened his first water-powered factory equipped with a power loom in 1771 and followed with improvements in carding and roving (1775). These developments signaled the practical beginning of an Industrial Revolution in which machines supplemented and then replaced handwork. This equipment represents the first stage of the development of machinery that would manufacture the hand tools that were still laboriously made by individual craftsmen and blacksmiths throughout England and America.

A second generation of English Industrial Revolutionaries lead by Henry Maudslay (1771 - 1831) were equally significant in terms of their later impact on hand tool production. Maudslay may have been the most innovative machine designer in the history of the Industrial Revolution. The power tools that he built and the engineers whom he trained or influenced (James Nasmyth, Joseph Clement, Joseph Whitworth, etc.) played a key role in substituting machine-made machines for the craft-based technologies that ironically lasted longer in England than America. Maudslay was preceded by John Wilkinson, who invented the boring machine necessary to hollow the cylindrical cavities of Watt's steam engine pressure vessels. John Jacob Holtzapffel began his family's long tradition of exquisite lathe-making in London in 1787. Maudslay, who first worked at the Woolwich arsenal in London at age 12, became a metalsmith by the age of 15 (1786) and was a well known toolmaker by 1799 (Cantrell 2002). During the next 30 years, he invented table engines, compound slide rests, self-tightening collars for hydraulic presses, and spring-winding machines, to mention only a few of the inventions that were key elements in the industrialization of England. Maudslay designed stationary steam engines and marine engines, sawmill machinery, gun-boring machinery, and hydraulic presses

7   *A view over the Coalbrookdale Upper Works by Francois Vivares, dated 1758 and packed with interesting detail. On the left is an engine cylinder being transported from the boring mill further down the Dale. The buildings in the middle include the tall chimneys of two air furnaces and the wider top of the blast furnace belching smoke. To the right is the reservoir for the waterwheels, with the coke hearths in the foreground. In the background beyond the reservoir are Dale House and Rosehill House, with Sunniside further up on the horizon, all ironmasters' houses, and Tea Kettle Row on the side of the hill with its characteristic dormer windows*

Figure 34. Richard Hayman and Wendy Horton, 2003, *Ironbridge: History & guide*, Tempus Publishing Ltd., Stroud, Gloucestershire, UK, pg. 23. Reprinted with the permission of Ironbridge Gorge Museum Trust - Elton Collection - AE185.769. Of particular importance in this view of Coalbrookdale is the depiction of John Wilkinson's engine cylinder being transported by horse drawn wagon to some undisclosed location. This engine cylinder, which facilitated the efficient functioning of Watt's steam engine, is a symbol of the coming age of the factory system of mass production of guns, tools, and machinery with interchangeable parts.

that insured his fame as the leading innovative design engineer of the early 19th century. His standardized screw threads were precursors to the later work of Joseph Whitworth. Maudslay's 1797 improvements to the screw-cutting lathe alone would have insured his fame. Between 1802 and 1807, with the help of the French émigré Mark Brunel, he designed and constructed 45 machines for the mass production of ships' blocks for the British Navy (Cooper 1984). The modern age was off to a running start.

Another of the innovators of the machine age, Richard Roberts (1789-1864) was perhaps most famous for his design of the London Bridge. Working for Maudslay before the end of the Napoleonic wars in 1815, Roberts made important improvements to the power loom, spinning mule, slide lathe, planing machine, and he was manufacturing gear-cutting engines at Manchester by 1830. Roberts also produced important slotting machines and a punching and shearing machine. These all played a role in supplementing handwork and producing the machinery that would eventually result in the rise of the factory system for all tool production after the mid-19th century in America and then later in England.

David Napier (1788-1873) specialized in manufacturing printing presses. His inventions and innovations dominate printing press design until the advent of photo composition in the 1970s (Cantrell 2002). He worked for Maudslay, later designing a tracing machine, a bullet-making machine, coin-sorting equipment, and a registering compass. Joseph Clement, (1779-1844) who was Maudslay's chief draftsmen and chief marine engine designer, was a close friend of Napier and invented an important ellipse machine for accurate perspective, as well as turning lathes, the barrel tail stock, and milling cutters. Clement also produced the first taper, middle, and plug taps for Maudslay in an environment where standard measurements for threads had still not been widely adopted (1830). All was soon to change under the influence of both American innovators and Joseph Whitworth, whom Clement clearly influenced.

Joseph Whitworth (1803-1887) was another engineer who worked under Maudslay and then emerged as one of England's most important machine designers. He obtained 48 British patents and is probably best known for his standardized screw thread measuring systems and his use of decimal, rather than fractional, measurements in the precision tools that he designed. Whitworth left Maudslay's employment in 1828, and, after he worked with Joseph Clement, he moved to Manchester in 1832 and became one of England's most successful industrialists. His innovations in precision measurement were quickly adapted in the United States. Occasional surviving examples of his cylindrical gauge measuring set, his hand screws, radial drilling and boring machines, and other "J. Whitworth, Manchester" signed equipment are still to be found in old workshops across the United States. His precision measuring machine was reliable to within millionths of an inch. Whitworth was part of the English Royal Commission that visited New York in 1853 to evaluate America's rapid advance in mass production techniques using standardized parts.

Joseph Nasmyth (b. 1809) was another Maudslay-trained engineer and leader in machine tool technology innovation. He designed a road steam carriage, high pressure steam engine, which recycled waste steam to increase engine efficiency, and he invented the flexible shaft to transmit rotary motion. Nasmyth's nut-cutting machine (c. 1841) was one of the early forms of a milling machine. Nasmyth is best known for patenting his steam hammer in 1842 and manufacturing it from 1843 onward. The efficiency of this hammer greatly facilitated industrial production of heavy equipment, especially railroad locomotives, allowing Nasmyth to establish one of the largest industrial complexes in England at Manchester, i.e. the Bridgewater Foundry. Another engineer influenced by Maudslay and his followers was William Muir, who invented the letter press. Many other inventors and engineers of lesser fame were influenced by Maudslay, but his legacy as the most significant of the English industrial revolutionaries overshadows the reputation of all other English engineers (Cantrell 2002).

Maudslay had first achieved significant renown working with Joseph Bramah in London, helping Bramah produce his patent lock (c. 1790-1797). Maudslay then went on to become the leader of an engineering cartel who designed the machines that forever changed the nature of manufacturing, including hand tool production, during the classic period of the Industrial Revolution. He was the leading innovator in the development of the factory system of production using interchangeable parts manufactured with modern drop or die forging techniques. The demand for increasing quantities of steel as a result of these innovations gave rise to the bulk steel manufacturing processes of the second half of the 19th century. This third stage in the Industrial Revolution foreshadowed the end of crucible steel production made from high quality Swedish charcoal iron, which had been converted to blister steel in England's ubiquitous cementation steel furnaces before being broken into small pieces and remelted in clay crucibles. By 1910, crucible steel production was primarily used to make alloy, including high speed, steels, rather than cast steel for edge tools, which were no longer in significant demand. The demise of crucible steel production occurred during the 1920s as the electric arc furnace replaced the crucible process for alloy steel production. A rapid and significant decline in the quality of steel edge tools soon followed the end of crucible cast steel production.

The greatest irony in the florescence of the English engineering pioneers, was that, while they designed and built the machinery of the Industrial Revolution, American entrepreneurs and craftsmen quickly adapted and improved their designs and incorporated them in a factory system of mass production that soon left most English hand tool manufacturers, often still using craft-based technologies, unable to compete with the more efficient American factories. While the English industrial revolutionaries were laying the groundwork for mass production of machinery, firearms, and tools with interchangeable parts, which was the fundamental characteristic of the American factory system, other English inventors, metallurgists, scientists, and eccentrics were preparing the way for that other essential ingredient of the factory system of mass production, i.e. bulk steelmaking processes.

Edge Tools, High Speed Steel, and Metallography

When Benjamin Huntsman adapted the ancient strategy of crucible steel production to his watch-spring-making business and then started producing cast steel coin dies and cast steel for other customers, he became the most famous facilitator of the art of edge toolmaking. The birth of the sciences of metallurgical chemistry and metallography were a century in the future despite the growing curiosity about and knowledge of the physical world that characterized the Enlightenment. The strategies and techniques of steel and edge toolmaking were galloping way ahead of the snail's-paced advances in the understanding of their chemistry and microstructure. By the time the sciences of metallurgy and metallography had caught up with modern steelmaking technologies, perhaps best symbolized by Bain's (1939) publication of *Functions of the Alloying Elements in Steel*, that now obscure branch of metallurgy, the art of forging edge tools, was in rapid decline.

One of the most interesting, if not puzzling, footnotes to the history of metallurgy and the rise of industrial society is the fact that the manufacture of the highest quality edge tools ever made occurred well before the metallurgy and metallography of their composition was well understood. Once Huntsman perfected the art of casting steel in those Stourbridge clay crucibles (1742 – 1750), the highest quality cast steel woodworking tools were made first in Sheffield and Birmingham, England, for the next 150 years and then, after 1850, in the river valley towns of New England.

First made by multitasking blacksmiths and shipsmiths, the edge tools of the shipwright, were "tool steel" of the most basic iron-carbon mixture. The definition of tool steel is "iron containing >0.5% carbon." The 0.5% carbon content is the practical cutoff point below which steel cannot be significantly hardened by quenching (rapid cooling). As produced by the Bessemer process and similar to malleable iron in its carbon content, mild steel cannot be hardened by quenching nor made into useful edge tools by subsequent tempering to relieve brittleness and restore some degree of ductility to its cutting edge. Blacksmiths knew the physical difference between wrought and malleable iron and steel; the chemical composition of wrought iron, steel, and cast iron, and the role of their carbon content, was not understood until the late 18th century. It was not understood until the mid-19th century that microconstituents, such as silicon and manganese typically present in malleable iron in amounts less than 0.5% influenced edge tool quality, as did the microcontaminants phosphorus and sulfur. By the time the chemical composition and the wide variety of the microstructures in iron and steel were finally being deciphered, the former more quickly than the latter, the age of the edge tool and the wooden ships they built was coming to an end.

Over a century after Huntsman's reinvention of cast steel, Robert Mushet created his Self Hard tungsten-manganese alloy tool steel, the purpose of which was not to build wooden ships, but to efficiently and effectively cut and shape, on lathes and milling machines, the malleable cast iron and steel components of the machines that would be the basic components of the third stage of the Industrial Revolution. The worldwide spread of the American factory system of drop-forged interchangeable parts soon followed.

Just after Robert Mushet helped Henry Bessemer perfect the bulk process of converting cast iron to malleable iron (low carbon steel), he invented the dozzle for eliminating the pipes (cavities) in ever larger cast steel crucibles (1862). In this year, he also rebuilt his isolated Titanic Steel and Iron Company in the Forest of Dean and began producing alloy steels. If properly forged and heat-treated, the simple constituents of tool steel used by woodworkers, i.e. carbon and iron, could be made into edge tools of the highest quality but were inadequate for cutting or milling the wide variety of malleable cast irons, and, later, mild, then alloy steels, that constituted the machinery of the Industrial Revolution. Edge tools may have built the wood frames of the first lathes, table saws, and drill presses, but Mushet's Self Hard alloy steel cutting tools were a key component of the spread of industrial society. The discovery that tungsten and manganese, when combined with that iron-carbon alloy, steel, would make iron and steel cutting tools that could withstand high temperatures and be used at high speeds on lathes and milling machines (1859) was followed by the gradual inclusion of other elemental metals in the tool steel family of alloys as follows: chromium then molybdenum, vanadium, and cobalt. By the beginning of the 20th century, the art of high speed steel cutting tool fabrication had been perfected. Dozens, then hundreds, of special purpose high speed steels were being produced by the mid-20th century. High quality woodworking tools continued to be manufactured out of by then old fashioned cast steel, and also, possibly, out of steel produced in the electric arc furnace by a few companies, such as the James Swan Company of Seymour, CT (closed 1949). Until the age of the electric arc furnace (1920), crucible steel production in quantities that were still only made in 25 – 50 kg batches continued but only for special purpose steels, the most important of which were alloy steels for cutting tools. The huge capacity (25 - 50 ton) furnaces of the Bessemer pneumatic and open-hearth processes met the routine needs of an industrial society now building automobile factories and skyscrapers. Edge tool production was now a forgotten footnote in the age of the internal combustion engine.

While Mushet was experimenting with alloy steel production in his obscure Forest of Dean facility, Henry Clifton Sorby began his historic investigation of the molecular constituents of metals. A lifelong resident of Sheffield and a wealthy member of a famed family of Sheffield cutlers, Sorby had the resources and leisure time to investigate, with the help of recent advances in photomicrography, the microstructures of minerals. In

1863, he obtained thin samples of Bessemer steel, etched them, and then produced photomicrographs of Bessemer's product. In 1863-4 he made extensive investigations of all ferrous metals, especially including the re-crystallization patterns that resulted from annealing. Sorby's photomicrographs of cast iron, steel, and wrought iron were the first accurate depictions of the differing microstructures of these ferrous metals. Sorby's research, first presented as an exhibition of etched samples at Newcastle-On-Tyne in 1863 and in lecture format in 1864, was, unfortunately, never fully published or widely known until his release of an 1885 paper and its full publication in 1887, which included illustrations made with the help of the newly invented Woodburytype printing process (Smith 1960, 167-85). In the intervening period between 1863 and 1887, vast advances in the understanding of the chemistry of metallurgy had occurred. Ironically, Sorby, whom Cyril Smith (1960) calls the "father of metallography," did not see widespread circulation of his landmark discoveries on the microstructure of ferrous metals until after the publication of his final paper. After that date, vast advances occurred in unraveling the microstructures of all ferrous metals, a science finally synthesized by Edgar Bain in 1939.

Almost four millennia of the forge welding of iron tools and weapons occurred before a scientific understanding of the chemistry, alloy content, and metallography of ironmongering emerged in the 19th century. During this period of time, the legacy of technique was the fundamental criteria for the successful forging of edge tools and weapons. This legacy was the essence of the "kan of the ironmonger," a combination of the ritual of "rule of thumb" bloomsmithing and forging practices and the empirical experience of centuries of forging. Typical of the empirical experience of forging hot iron was the gradual discovery that quenching at a high temperature resulted in a hardening of steel. It was not known that this hardening suppressed the microstructural decomposition of the steel object being forged. More obvious as a central component of empirical experience was the understanding that hammering, especially in conjunction with reheating, restored the strength of an edge tool and lessened its brittleness. Also unknown were the molecular changes involved in the re-crystallization of ferrous metals to a stable equilibrium, which resulted from tempering and hammering (Smith 1960).

One of the ironies of the Industrial Revolution and the spread of industrial society across American and other landscapes was the use of handmade hand tools and files to construct and shape the first machines that were used to drop-forge hand tools, firearms, and other equipment, often with interchangeable parts. 1827 is a landmark date in the evolution of the American factory system, marking the first full year of the operation of the Samuel Collins Axe Factory in Collinsville, CT, Collins and his toolmakers, including Elisha Root, etc., quickly designed equipment to shape and form hand tools, such as axes. During the late 1830s and 1840s, an increasing number of hand tools were machine-made, eventually being used to construct machinery that would make hand tools obsolete, at least until the collapse of non-sustainable industrial society and the advent of a post-apocalypse creative economy, now imminent, that may reverse the gradual decline in the use of hand tools. The late 20th century substitution of computerized programs for the use of the micrometer is an example of one more step in the gradual obsolescence of hand tools, an ironic prelude to the birth of a new age of convivial hand tools.

Railroads and the Bessemer Process

In the period after 1840 and especially after 1860, the most significant source of the growing demand for iron and steel was not warfare, but railroads. The rapid spread of railroads, whether to Sheffield steel-producing furnaces or from Boston to Worcester, MA, the site of the Coes wrench factories, was the key to rapid industrial growth in England, continental Europe, and America. Huge quantities of cast, malleable, and wrought iron were fortuitously produced by recently improved steam-powered blast furnaces and the now ubiquitous reverbatory furnace. But cast iron was useless for railroad rails and inadequate for many other essential applications. Wrought and malleable iron were widely used but their inherent lack of strength limited the size and durability of engineering projects ranging from rails to bridge, ship, and building construction. During this time period the search for innovative strategies for increasing steel production became intensified. Henry Bessemer, a most unlikely figure with no experience in metallurgy or chemistry emerged and, with a stroke of good luck, solved at least part of the problem of the need for more steel.

Henry Bessemer was not a metallurgist, but an inventor who filed 116 patents between 1838 and 1883, ranging from the production of telescope lenses and velvet, to the refining of sugar and improved production of ordnance from wrought iron. During experiments with the later project, Bessemer made one of the most significant observations of 19th century industrial history. While blowing air into a puddling furnace to speed up the production of wrought iron by the oxidation of malleable iron, he noted the complete decarburization of scraps of pig iron on the edge of the furnace lining (Barraclough 1984b, 113). Bessemer then began directing air blasts through small quantities of liquid cast iron, quickly and efficiently producing slag-free, low-carbon iron in a 20 kg crucible charge. Bessemer soon experimented with larger quantities of pig iron in a converter producing hundreds of kilograms of malleable iron of a relatively high quality. While doing these experiments, the gods must have favored Henry Bessemer, who had the initial good luck of re-melting cast iron made from the low phosphorus hematite ores of the Forest of Dean, which also contained significant traces of manganese, that anti-oxidizing agent, which played such an important role in the production of natural steel in Noricum and, later, in the fining of manganese-laced pig iron to produce German steel.

Bessemer quickly announced his success to the steel-producing industries that were anxiously searching for new strategies for the bulk production of steel. August 13, 1856 is the date on which Bessemer read his famed paper "The Manufacture of Malleable Iron without Fuel" at a meeting of British industrialists at Cheltenham. Ironically, Bessemer's attention focused on the efficient production of malleable iron. At this time, he had no idea that his innovation would revolutionize steel production technology. The steel producers quickly realized that Bessemer's innovative pneumatic process for the rapid oxidation of pig iron using an air blast could also be used to manufacture steel, and three English steelmakers and a Welch tinplate manufacturer immediately bought the rights to his process. Unfortunately for Bessemer, all licensees initially failed to produce a forgeable, useful product. All, including Bessemer, a non-metallurgist, were ignorant of the fact that they were using pig iron with a relatively high phosphorus content and were also unaware of the important role of manganese containing spiegeleisen in preventing over-oxidation. Phosphorus embrittles iron in the presence of carbon, making it useless. The hot air blast also over-oxidized the malleable iron without the presence of manganese, and Bessemer had to return all his royalties to his first licensees.

With the help of the highly regarded chemist and steel historian John Percy, Bessemer spent several years attempting to solve the problem of excess phosphorus in his ores, but failed yet again, due to the use of a flux contaminated with phosphorus. Aside from his creative inventiveness, Bessemer had one other resource to mitigate his metallurgical and chemical ignorance, and that was Robert F. Mushet, an obscure metallurgist with a nearly clandestine steel furnace in the Forest of Dean north of the Severn River. Mushet was a friend of Bessemer and well aware of his initial research, his 1856 paper, and the subsequent failure of the attempt to produce malleable iron rapidly by the pneumatic air blast process. Working in the obscurity of the Forest of Dean, Mushet realized that Bessemer's iron was overly oxidized and recarburized it by re-melting it and adding spiegeleisen. The combination of Dr. Percy's knowledge of the sources and detrimental role of phosphorus in steel production and Mushet's metallurgical finesse and knowledge of the importance of the role of spiegeleisen in steel production enabled Bessemer to solve the problems blocking successful production of low-carbon steel.

Unfortunately, he had no more licensees willing to pay him royalties to use his process, and, in 1856, he began constructing his own steelmaking facility in Sheffield, which was completed and in operation in 1859. In the meantime, another early licensee, Goran Frederik Goransson from Sweden, who had access to Sweden's low phosphorus charcoal-fired pig iron and made changes to the tuyère design, was the first to use the pneumatic process successfully to produce steel on July 15, 1858. Goransson's Swedish pig iron was also high in manganese, and Goransson sent 15 tons of ingots to Bessemer at his new Sheffield facility, confirming both the quality of the steel and the success of the process (Barraclough 1984b, 115).

Between 1860 and 1862 a number of Sheffield steelmakers obtained licenses from Bessemer after observing the success of his new plant, and Bessemer converters were soon in use at Sheffield and other locations, principally to make steel rails for England's rapidly expanding network of railroads. In particular, the community of Barrow, located adjacent to one of England's few remaining low phosphorus hematite ore deposits quickly emerged as one of England's major steel-producing facilities.

> From a vacant site in 1863, there were 11 blast furnaces, each producing 4500 tons of 'Bessemer pig' per year, by 1867. By 1872 it had six 5 ton converters and twelve 7 ton converters; by 1880 it was annually producing 150000 tons of steel, about one-eighth of the British total, and the population of the town had quadrupled in 17 years. It was the largest steelmaking plant in the world for a short period. (Barraclough 1984b, 116)

While Bessemer was perfecting the pneumatic process for bulk steel production, the Martin Brothers in France and William Siemens and his brother Frederick were developing an alternative method of bulk steel production that would eventually supercede the pneumatic process for most bulk steel needs.

The Siemens-Martin Open-Hearth Furnace

The evolution of the regenerative open-hearth furnace was concurrent with Bessemer's development of the pneumatic process, where liquid cast iron was quickly decarburized in ±20 minutes into malleable iron (i.e. low-carbon steel). In any regenerative gas furnace, the hot gasses produced in the smelting or melting process were captured and recycled, heating up regenerators at one end of the furnace, which would then preheat both incoming air and furnace gasses in a reversing pattern for about 20 minutes prior to their being mixed together for combustion in the furnace interior. It was significant that the regenerator furnace, with its attached regeneration chambers, could achieve more efficient smelting or melting with a hotter flame, saving time and fuel. The innovative open-hearth regenerative hearth furnace was designed by William Siemens, but was originally invented by his brother Frederick and patented in 1856. Its original aim was to reheat iron and steel billets prior to re-rolling. The furnace was soon used for steelmaking, copying the ancient Chinese and the later Italian Renaissance Brescian strategy of melting high-carbon cast iron and low-carbon malleable iron in appropriate amounts in an open-hearth furnace according to the type of steel desired. The first regenerative melting furnace designed by Siemens operated in England in 1862 and was quickly followed by a more elaborate process using a combination of cast iron, scrap iron, and scrap steel, a strategy devised by two brothers, Pierre and Emile Martin, in a Siemens furnace built at Seraing, France.

Initially designed as a continuous process and later adapted to batch steel production, the Siemens-Martin open-hearth furnace used the protective cover of the liquid slag to keep the liquid iron-steel mix from being oxidized by furnace gasses. The melting process took 4 – 6 hours compared to the 20 minute operation of the Bessemer pneumatic process, and the melted steel was then poured into revolving molds. The substitution of puddled iron for steel scrap or puddled steel blocks in the open-hearth process could produce low-carbon steel, which could not be hardened by quenching due to its low-carbon content, otherwise a higher carbon steel was produced. In England, the Siemens brothers took out patents in 1864 and 1865. The Siemens-Martin process, also known as the "pig and scrap" process, operated briefly as an important bulk steel-producing strategy before being replaced by the Siemens open-hearth process. The Siemens-Martin process was strictly a melting process, the carbon content of the product being determined by the carbon content of its initial ingredients. The efficiency and economy of the regenerative gas furnace and the large capacity of the Siemens-Martin open-hearth furnace made it, along with the Bessemer pneumatic process, one of the first, if short-lived, bulk steel production processes.

The Siemens Open-Hearth Furnace

The Siemens-Martin open-hearth pig and scrap furnace had a major disadvantage that it shared with the initial production strategies of the Bessemer process. In 1862, routine chemical analysis of the carbon content and alloy microconstituents of scrap metal was not yet part of the steelmaking regimen. Quality control of both procedures had not been established. Additions of carefully measured amounts of spiegeleisen at the end of the melt would soon greatly improve the Bessemer process. The Siemens open-hearth process differed from the Siemens-Martin process, which it replaced, by the addition of measured amounts of iron ore to the slag after the melt was underway. The iron oxide content of the slag was increased, resulting in spectacular boiling reactions caused by the burning of carbon monoxide produced at the slag-metal interface as a result of the reaction of iron oxide with the carbon. Barraclough (1984b) provides a handy synopsis of the resulting reaction for the non-chemist.

The reactions may be expressed as follows:

$$2Fe + O_2 = 2FeO$$

$$FeO + C = Fe + CO \text{ (gas)}$$

At the same time the manganese and silicon are oxidized:

$$Si + 2FeO = SiO_2 + 2Fe$$

$$Mn + FeO = MnO + Fe$$

From this silica so produced, plus further silica dissolved from the furnace lining, a liquid slag would also be formed, consisting of a mixture of silicates, which in their simplest form may be represented as follows:

$$2MnO + SiO_2 + 2FE$$

$$2FeO + SiO_2 = Fe_2SiO_4$$

Meanwhile, the carbon monoxide brought into contact with air burns with a blue flame:

$$2CO + O_2 = 2CO_2$$

In the Siemens process, the reactions are essentially the same, except that the source of the iron oxide is the added iron ore rather than the oxidation of some iron by the oxygen in the air, as in the Bessemer process. In the Bessemer process, the process is cyclical: iron is oxidized, the iron oxide reacts with carbon, etc., producing the oxide of the other elements and releasing the iron again as metal, the net effect being the reaction of the oxygen from the air with the carbon, silicon or manganese. There is, however, always some loss of iron to the slag. In the Siemens process, the iron oxide is added as such and there is therefore additional metallic iron produced as the other elements are removed. (Barraclough 1984b, 110)

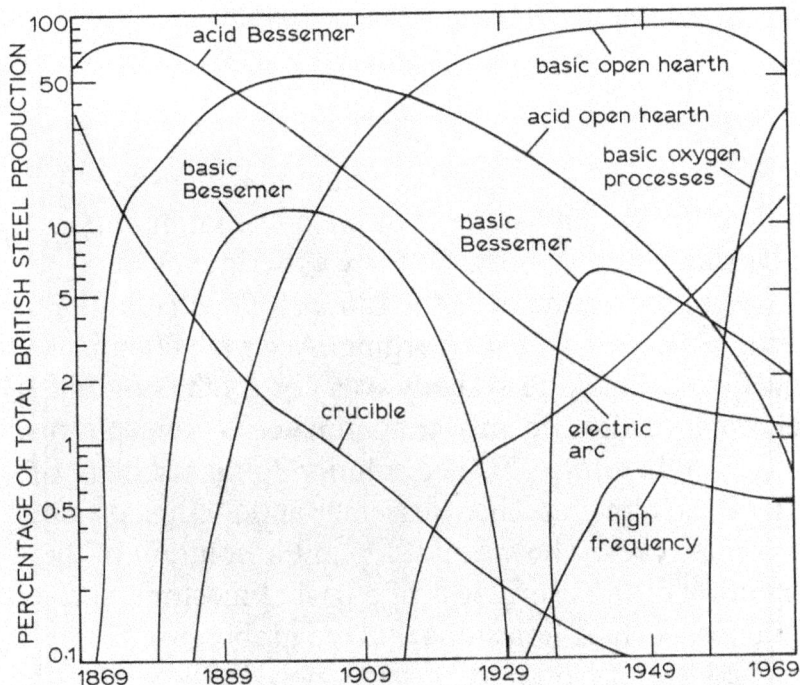

Fig. 9   British steel production, 1869–1969, show-
ing the relative importance of the various
processes

Figure 35. British steel production. Barraclough, K.C. (1984b).
Steelmaking before Bessemer: Crucible steel the growth of technology.
Volume 2. London: The Metals Society. pg. 108, Fig. 9. Used with the
permission of the Metals Society.

Barraclough (1984b) then explains the basis for the longevity of the Siemens open-hearth process. It allowed for significant quality control as a result of the ability of the furnace operator to determine and control slag composition and thus it's oxidizing capability. More iron ore increased oxidation; oxidation could be slowed by adding limestone and manganese oxide. Towards the end of the boil, "spiegeleisen could be made to stop the boil entirely" (Barraclough 1984b, 111). Ferrosilicon (45% Si, 55% Fe) and ferromanganese (80% M, 5% C, 15% Fe) could also stop the reaction, after which carefully controlled additional amounts of pig iron and/or other carbonaceous materials could be added to adjust the carbon content. Acid open-hearth followed by basic open-hearth bulk steel production, along with acid and basic Bessemer processes, dominated British and world steel production from 1865 until the appearance of the electric arc after 1910 and then the basic oxygen processes after 1950. By 1970, these two processes equaled, then surpassed, the basic open-hearth process in total steel production.

Acid versus Basic Steelmaking

A fundamental problem bedeviling all steel production strategies from the beginning of the Iron Age to 1879 was the uptake or absorption of both sulfur and phosphorus from iron ores, furnace linings, and, in the case of sulfur, from coke and coal fuels, into the iron and steel being smelted. Sulfur contamination in steel makes it "hot short." On hammering, hot short steel disintegrates. The worst case scenario for sulfur contamination would be the equivalent of hammering a tomato. Phosphorus creates "cold shortness" making iron or steel containing any significant amount of carbon difficult to forge at any temperature, especially at room temperature.

One of the two problems that nearly sabotaged Bessemer's attempt to make steel by the pneumatic process was the contamination with phosphorus of the ores that his first licensees used and contamination of slag by phosphorus in early experiments once he reverted to using low phosphorus ore. In the period of 1858-1865, an understanding of the chemical vagaries of the smelting process was just emerging. Accurate chemical analysis of the constituents used in any steelmaking strategy was just on the horizon. The majority of the earth's surface iron ores are significantly contaminated by phosphorus, making them useless for steel production before 1879. The solution to the problem of phosphorus came in three stages. First, steelmakers like Bessemer and Siemens soon learned to avoid high phosphorus content ores. The second step in the solution of the phosphorus problem was the recognition of the usefulness of a basic limestone-derived slag to absorb phosphorus. Basic fluxes, such as limestone and the grabbo used at Hammersmith (Saugus Ironworks) had been favored by furnace-men and bloomsmiths since the early Iron Age, but nobody understood the chemistry of their usefulness. A more scientific understanding of the chemical basis for using limestone as a flux evolved in the early 19th century. The practical application of this knowledge was developed in Bohemia in 1855 but did not become widely accepted until the late 1860s. The third obstacle, which remained unrecognized, or at least unresolved, until the 1870s was the necessity of replacing acid furnace linings, usually sandstone, brick, or other silica-containing materials with a less reactive refractory material. Famous steelmakers and metallurgists, such as Krupp in Germany and Lowthem Bell in England, had been unsuccessful in solving this problem. In 1877, Sidney Gilchrist Thomas, a police clerk taking a chemistry course, learned about the phosphorus-in-steel problem. Working with a cousin who was a steelworks chemist, Thomas succeeded in solving the problem with a combination of limestone bricks as refractory linings and lime additions to the Bessemer converting process. Thomas wrote a paper for the September 1878 meeting of the Iron and Steel Institute in Paris, but it was not read. However, an English steelmaker, Windsor Richards from Middleborough, read Thomas' paper and then provided the two cousins with converters and brick-making equipment. The lining blocks they made with "hard fired dolomite brushed with coal tar" (Barraclough 1984b, 122) facilitated the presentation of Thomas' paper in May of 1879, a final chapter in the rise of bulk steelmaking by pyrotechnic western society. As Barraclough (1984b) notes, both the basic Bessemer process and the basic open-hearth process led to the rapid expansion of steelmaking in continental Europe and in the United States, both of which had the majority of their iron deposits laced with significant quantities of phosphorus. The basic lining/basic slag open-hearth steel production technique became the most common steel-producing strategy from 1910 – 1970.

The net effect of the development of the Bessemer pneumatic, Siemens open-hearth, and basic steelmaking processes was the facilitation of mass production of low carbon steel for the massive construction projects of late 19th century industrial society. Whole cities of bulk process (modern) cast steel framed buildings, subways, factories, and railway stations would then be constructed. It is almost a forgotten footnote to the Bessemer pneumatic and Siemens open-hearth process that these bulk process steel products also facilitated mass production of drop-forged tools. Successful mass production of drop-forged tools, which initially began with hundreds of wrench makers, for example, drop-forging their uniquely designed tools, depended on one further scientific development, i.e. the identification and incorporation of steel alloys, such as tungsten, magnesium, chromium, and vanadium in the tool-forging process. This rather sudden late Victorian understanding of the chemical basis for successful alloy steel production allowed hundreds of small American wrench-makers, for example, to begin mass production, relatively speaking, of the many wrench forms sought by collectors today and documented by Cope (1999).

Alloy and High Speed Steels

When the Englishman Benjamin Huntsman rediscovered the art of crucible cast steel production in 1742, a most important historical event, the word "carbon" was unknown to the world. The science and chemistry of ferrous metallurgy was as unfathomable as the microstructures characteristic of the phase transformations from austenite to martensite. These are obscure inscrutable words unknown even today by most high school students and teachers. And yet, for 2,000 years every practicing ironmonger knew intuitively the art but not the science of forging the edge tools used to build the wooden ships that first sailed the Mediterranean and then the Atlantic Oceans. It was not known why those iron ores from Mount Erzberg, the "Iron Mountain" in Carinthia (Austria), were so much better for making the natural steel used to forge the short swords of the Roman legions. Manganese was another unknown word, but ironmongers could see and feel the superiority of steels made from ores mined at what were, essentially, sacred sites. The iron currency bars smelted in what is now Austria were transported down the "Iron Road" of central Austria to more northern Celtic and then Germanic metallurgists for almost two millennia. Two thousand years after the beginning of the florescence of Celtic metallurgy in central and eastern Europe, obscure European scientists (always European, never American) were beginning to elucidate the chemical and molecular structure of ferrous metals. Carbon, not phlogiston, was the key ingredient in baked iron as steel. Bergman, Rinman, Van der Monde, Berthollet, and Monge are obscure names in the history of metallurgy, but they were important contributors to the unraveling of the chemistry of ferrous metallurgy.

In the period between 1819 and 1825, Michael Faraday began searching for a new alloy to add to steel to reproduce "damascene steel" (Barraclough 1984b, 125). He had been preceded by Johann Conrad Fisher in Switzerland (1814) and was also competing with Berthier in France. Faraday experimented with nickel, chromium, copper, silver, gold, platinum, and rhodium in a fruitless search to find alloys to produce corrosion resistance in steel. In his two experiments with chromium, both he and Berthier anticipated the later success of Harry Brearley (1913) in fabricating stainless steel. These were the first tentative attempts to make alloy steel using Sheffield crucibles.

Almost 35 years later, Robert Mushet, who assisted Bessemer in perfecting the pneumatic process of bulk steel production by alerting him to the role of manganese, began experimenting with tungsten as an alloy of steel. He then set up his Titanic Steel and Iron Company at Darkhill in the Forest of Dean, using titanium as an alloy of carbon steel. Titanium produced no useful results, but Mushet also experimented with manganese, chromium, and tungsten. In 1868, he introduced the first important alloy tool steel, which was used for the next 30 years, the famous Self Hard steel (8% tungsten, 2% carbon, and 1% manganese) also known as R. M. S. for Robert Mushet Steel. Self Hard alloy steel was unique in its capacity to cut after being slowly air-cooled rather than quenched and tempered. Further heat treatment strategies, including its later adaptation by the American metallurgists Taylor and White, extended the usefulness of air-cooled alloy steel to other specialized tasks. In all of Mushet's alloy steel compositions, silicon and manganese, as well as the ubiquitous tungsten, were always present. By the late 19th century, many European and American companies were producing variations of Mushet's Self Hard alloy steel, despite the secrecy surrounding alloy steel production.

A final series of stages in the evolution of alloy steel technology included the development of high speed steels containing reduced carbon content. These were also characterized by lower silicon and manganese content, the continued use of chromium, and often greatly increased levels of tungsten ranging from 14 – 22%. The last stage in the evolution of high speed steels involved the addition of small quantities of vanadium and variable amounts of cobalt and molybdenum. The latter alloy may be the most important constituent of late 20th century alloy steels in the age of nuclear power plants and jet engine turbines, in which high temperature superplasticity is the critical element in providing the margin of safety required for their operation.

During the early years of the 20th century, Sheffield declined as an important steel-producing center, while Pittsburg, PA, achieved its temporary ascendancy as a steel production center for a nation that would undergo an increasingly rapid rate of industrial and economic decline in the later years of the 20th century. All these advances in alloy and high speed steel production were in the context of the ironic decline in the ability of all steel-producing communities to produce high quality edge tools. Since demand for woodworking edge tools had become minimal, there were no incentives for maintaining the high standards of the classic periods of English (1750 – 1890) or American (1840 - 1930) edge tool steel production. In the early 21st century we have come full circle, and the ability of local communities to make high quality edge tools could be an important component of sustainable creative economies, admittedly a pie-in-the-sky ideal in the age of rapid growth and potential collapse of a non-sustainable global military-industrial consumer economy.

A key to the success of the effort to revive the production of high quality woodworking edge tools and support the regional and local sustainable woodworking trades that would use these tools (timber framers, wooden boat shops, furniture makers, etc.) is a familiarity with the diverse history of steelmaking. The ancient roots of the legacy of the techniques used to forge edge tools are a key component of this history. The diversity of both steelmaking strategies and toolmaking techniques played an important role in the heyday of the wooden age. The legacy of the empirical experience underlying two millennia of wooden shipbuilding will be a key element in the revival of sustainable industries using renewable resources, including forest resources, which are already reappearing in formerly clear cut regions of New England.

The Legacy of Technique

In the United States, the last best practitioners of the art of making edge tools worked at the James Swan Company in Seymour, CT, which finally closed in 1951. One hundred years prior to the closing of the Swan factory, a robust community of toolmakers had

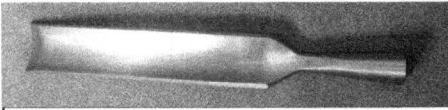

Figure 36. A socket gouge made by James Swan Co. of Seymour, CT. Cast steel, in the Davistown Museum IR collection ID# 40408DTM2.

evolved out of American colonial metallurgical traditions and industries that may never be fully documented. In this earlier era, the technical knowledge, but not the scientific understanding, of how to forge edge tools of the highest quality, was well established by the early 19th century, despite the wide variations in the quality of steel available for edge tool production. The toolmakers of New England and elsewhere, who were forging the slicks, adzes, mast shaves, and broad axes for what was, in the 1840s and 1850s, the zenith of the American wooden shipbuilding age, had the benefit of a millennium of the empirical experience of European sword cutlers, Solingen knife makers, and German Renaissance gunsmiths. These centuries of experience in the pattern welding of swords and knives and the metallurgy of fine guns, locks, and other malleable iron and steel artifacts were the basis for the evolution of a vigorous domestic edge toolmaking industry that arose in America in the 18th century.

Figure 37. Buck Brothers timber framing chisel. Cast steel, 16 ½" long, in the Davistown Museum MIV collection ID# 31908T20.

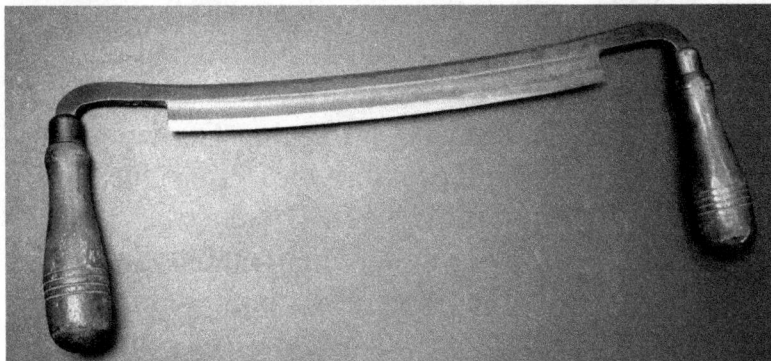

Figure 38. Thomas Witherby drawshave, 15 ¼" long, in the Davistown Museum IR collection ID# 31808SLP6.

When imported English cast steel bar stock became available to American edge toolmakers in the last half of the 18th century, American toolmakers quickly incorporated English cast steel to supplement sheaf and/or German steel to forge weld steel edge tools of the highest quality. The tools made with cast steel imported from Sheffield were always so marked to advertise their superiority. Now almost unnoticed, America's vigorous domestic edge toolmaking community was not only utilizing state of the art English cast steel, but continued to use what was probably carefully reforged blister steel, known as sheaf steel, to produce many of the larger sized timber framing, harvesting, and shipbuilding edge tools. Why domestic edge toolmakers could and did make such fine quality tools from steels other than English cast steel remains one of the mysteries of the success of the American edge toolmaking and shipbuilding industries in the years before domestic production of cast steel began (1860). The domestic production of the larger edge tools used by the shipwright, especially slicks, broad axes, mast shaves, and adzes, was accompanied by the ubiquitous and long enduring importation of fine Sheffield-forged carving tools, which played such an important role in the creative productivity of Salem, Boston, Newport, and Philadelphia case furniture makers.

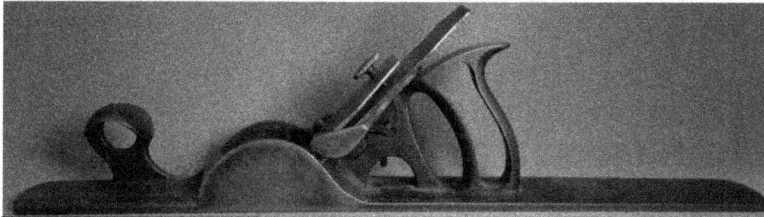

Figure 40. Birmingham Plane Co. jointer plane, one of the masterpieces of the classic period of American toolmaking. Malleable cast iron, 22" long, 2 ¾" wide, in the Davistown Museum IR collection ID# TJE4002.

While American planemakers before 1850 were almost entirely dependent on imported English cast steel blades, a vigorous metallic patented plane industry was evolving in New England mill towns that would soon capture the entire domestic market for the cast and malleable iron planes that were gradually replacing both the long established tradition of boat shop-made wooden razee planes and the prodigious production of birch, then beech, hand planes that had evolved in southeastern Massachusetts in the 18th century. By 1840, a vigorous, long-established domestic edge tool community was utilizing imported English cast steel for edge tool production. Of equal significance, as illustrated by the most important primary evidence of their existence, the tools themselves, domestic edge toolmakers, now long forgotten, working in obscure locations, such as Warren or Union, Maine, the Kennebec and Merrimac river watersheds or the Blackstone River Valley of Massachusetts, were also manufacturing edge tools of the highest quality from steel, which they had tediously repiled and reforged

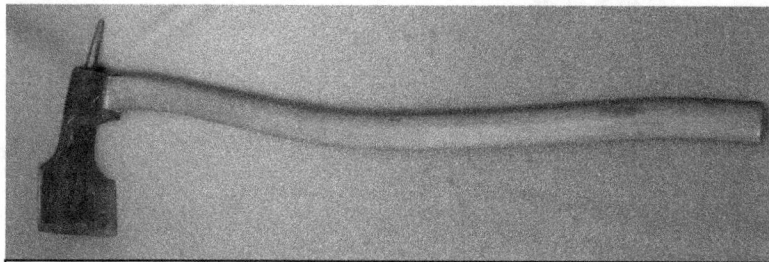

Figure 39. Classic Collins & Co., Hartford, CT, shipwrights' lipped adz. Cast steel, 10 ¾" long, and 5" wide with a 31" long handle, in the Davistown Museum IR collection ID# 62406T4.

from blister steel, sheaf steel (refined yet again), German steel, and possibly locally-produced natural steel. These tools, commonly encountered along with their cast steel counterparts even at the dawn of the 21st century in New England tool chests, collections, and workshops are not marked "cast steel" yet are often signed by their makers. It is these surviving edge tools that hold the secrets, many never to be re-narrated, of a vigorous New England-based domestic edge toolmaking community that had its roots in New England's first integrated ironworks at Saugus, MA, in 1646. Between this date and the perfection of edge toolmaking in New England in the mid-decades of the late 19th century by the Buck Brothers, Thomas Witherby, the Underhill clan, and others, hundreds of edge toolmakers were forging axes, drawknives, adzes, and timber framing tools in the riverine communities of New England, often using high quality steel they had forged or reforged themselves to supplement imported cast steel from England.

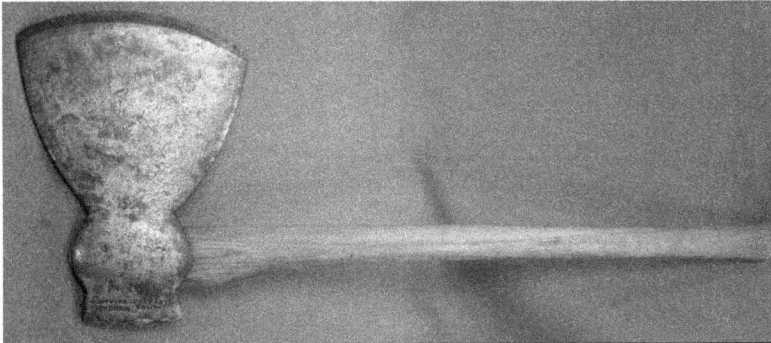

Figure 41. Broad ax made by C. Hunter, Bingham, Maine. Cast steel, 11" long, 7 ¾" wide with a 30" long handle, in the Davistown Museum MIII collection ID# 100605T3.

Between 1646 and 1846, vast improvements in the art of smelting and rolling malleable iron and steel and forging edge tools had occurred. Natural steel had long been supplanted by German steel. Blister steel became ubiquitous after 1700; the sources of sheaf steel forged or used by New England's edge toolmakers remains one of the unsolved archaeometallurgical mysteries of the era. Was it imported from England as was cast steel or was it made in America from reforged blister steel, which was produced in both countries? The appearance and ubiquitous presence of Sheffield made cast steel is well documented. The rise of bulk steelmaking technologies and alloy steels have since obscured the long history of steel- and toolmaking strategies and techniques that nonetheless were the basis for the rise of an American toolmaking empire that flourished from the early 19th century to its rapid demise in the late 20th century. Many chapters in the chronicle of this history remain to be compiled. The narration of the story of the roots and florescence of New England's unique community of shipsmiths and edge toolmakers from the early American colonial period to the florescence of American toolmaking in the classic period of the American Industrial Revolution is worth recording even if it is a forgotten historical footnote to the coming age of biocatastrophe. The following two volumes in the Davistown Museum *Hand Tools in History* series continue our exploration of the roots, labyrinths, and evolution of America's hand toolmaking industries in the context of 3,000 years of ferrous metallurgy.

Figure 42. C. J. Kimball & Son, NH, farriers' slick. Forged steel, 26 ¼" long, 1 1/3" wide, in the Davistown Museum MIV collection ID# 102904T8.

Appendix A: Definitions of Iron and Steel

Carbon content of ferrous metals: Sources vary widely in defining the *minimum* carbon content of steel, which ranges from 0.1 to 0.5% carbon. Please note the caveats that follow the definitions.*

Wrought iron: 0.01 – 0.08% carbon content (cc); soft, malleable, ductile, corrosion-resistant, and containing significant amounts of siliceous slag in bloomery produced wrought iron, with less slag in blast-furnace-derived, puddled wrought iron. Wrought iron is often noted as having ≤ 0.03% carbon content.

Malleable iron 1): 0.08 – 0.2% carbon content (cc); malleable and ductile, but harder and more durable than wrought iron; also containing significant amounts of siliceous slag in bloomery produced malleable iron, with less slag in blast-furnace-derived, puddled malleable iron.

Malleable iron 2): > 0.2 – 0.5% carbon content (cc). Prior to the advent of bulk-processed low carbon steel (1870), iron containing the same amount of carbon as today's "low carbon steel" (see below) was called "malleable iron." Its siliceous slag content gave it toughness and ductility, qualities not present in modern low carbon steel, hence its name. Before 1870, a wide variety of common hand and garden tools and hardware were made from malleable iron with a significantly higher carbon content than wrought iron.

Natural steel: 0.2% carbon content (cc) or greater. Natural steel containing less than 0.5% cc is synonymous with the term malleable iron. Natural steel is produced only by direct process bloomery smelting and was the predominant form of steel produced in Europe from the early Iron Age to the appearance of the blast furnace (1350). Small quantities of natural steel continued to be produced by bloomsmiths, especially in the bog iron furnaces of colonial New England, until the late 18th century, as well as in rural areas of Appalachia and southern Europe until the late 19th century.

German steel: 0.2% carbon content (cc) or greater. Steel made from the decarburizing of cast iron in finery furnaces. The strategy of making German steel dominated European steel production between 1400 and the advent of bulk process steel technologies, hence the term "continental method" as an alternative name for this type of steel production.

Wrought steel: 0.2 – 0.5% carbon content (cc); another name for malleable iron. Wrought steel was made from iron bar stock and was deliberately carburized during the fining process to make steel tools that are still commonplace today, such as the ubiquitous blacksmith's leg vise.

Low carbon steel: 0.2 – 0.5% carbon content (cc). Less malleable and ductile than wrought and malleable iron due to its lack of ferrosilicate, low carbon steel is harder and more durable than either and can be only slightly hardened by quenching. Some recent authors (Sherby 1995a) define low carbon steel as having 0.1% cc. Produced after 1870 as bulk process steel (e.g. by the Bessemer process), low carbon steel has all its siliceous slag content removed by oxidation before being strengthened by the addition of manganese and other alloys. Before the advent of bulk process steel production, there was no such term as "low carbon steel." All iron that could not be hardened by quenching (< 0.5% cc) was known as "malleable" iron, more recently often referred to as "wrought" iron.

Tool steel: 0.5 – 2.0% carbon content (cc). Tool steel has the unique characteristic that it can be hardened by quenching, which then requires tempering to alleviate its brittleness. Increasing carbon content decreases the malleability of steel. If containing >1.5% carbon content, steel is not malleable, and, thus, not forgeable at any temperature. Such steel is now called ultra high carbon steel (UHCS). Palmer, in *Tool Steel Simplified*, provides this generic description of tool steel: "Any steel that is used for the working parts of tools" (Palmer 1937, 10). All modern steels are characterized by the addition of alloys, including nickel, manganese, tungsten, chromium, cobalt, vanadium, and molybdenum for function-specific applications.

Ultra high carbon steel (UHCS): 1.5 – 2.5% carbon content (cc); a modern form of hardened steel characterized by superplasticity at high temperatures and used in industrial applications, such as jet engine turbine manufacturing, where extreme strength, durability, and exact alloy content are necessary. Powdered metallurgy technology is frequently used to make UHCS.

Cast iron: 2.0 – 4.5% carbon content (cc); hard and brittle; not machinable unless annealed to produce malleable cast iron. Cast iron comes in many forms, the following of which are the most important:

	Silicon %	Graphite % (Free Carbon)	Combined Carbon %	Total Carbon %	Properties
White cast iron	0.70	0.10	2.65	2.75	Very hard
Annealed malleable iron	0.70	2.70	0.05	2.75	Machinable

	Silicon %	Graphite % (Free Carbon)	Combined Carbon %	Total Carbon %	Properties
Cast iron for chilled castings	1.00	1.00	2.00	3.00	Very hard
Semi-steel	1.75	2.80	0.40	3.20	Machinable
Gray cast iron	2.00	3.10	0.30	3.40	Machinable
Soft gray cast iron	2.50	3.50	0.15	3.45	Machinable

Table 3 (Spring 1917, 180)

*Caveats to carbon content of ferrous metals

- Both modern and antiquarian sources vary widely in their definitions of wrought iron, malleable iron, and steel. Modern sources variously define steel and/or low carbon steel as iron having a carbon content greater than 0.08%, 0.1%, 0.2%, and 0.3%.

- Before the advent of bulk process steel industries (1870), which produced huge quantities of low carbon steel that could have a carbon content in the range of 0.08 – 0.5%, iron having a carbon content of < 0.5% cc was called malleable iron. Other generic terms for iron that could not be hardened by quenching (> 0.5% cc) were bar iron, wrought iron, and merchant bar.

- The 1911 edition of the *Encyclopedia Britannica* defines wrought iron as containing less than 0.3% carbon, cast iron as having 2.2% or more carbon content, and steel as having an intermediate carbon content > 0.3% and < 2.2%.

- Gordon (1996) defines steel as having a carbon content > 0.2%. This cutoff point is probably the most appropriate to use in defining steel, but also poses a problem since most sources define wrought iron as having < 0.08% cc; therefore, leading to the confusion of iron with a carbon content > 0.08% but < 0.2% as being either wrought iron, low carbon steel, or an orphan form of undefined iron.

- In view of the long tradition of the use of the term malleable iron, this text resurrects the use of that term to cover this gray area of the carbon content of ferrous metals.

Bibliography: European Precedents and the Early Industrial Revolution

This bibliography is the first of several contained within the *Hand Tools in History* series. Extensive additional information about the history and science of ferrous metallurgy and European and American toolmakers is contained in volumes 7 and 8. Volume 11 contains a special topic bibliography on metallurgy. Other special topic bibliographies follow the general topic bibliography in volume 8. All bibliographies can be accessed online at www.davistownmuseum.org. Volume 30 of the Davistown Museum publication series, *The Complete Bibliographies*, is available upon request.

Adamowicz, Laurent. March 2007. Codes and symbols of European tools, part I. *The Chronicle*. 60:1-9.

- This article has an extensive list of references.

Alexander, John. 1962. Greeks, Italians and the earliest Balkan Iron Age. *Antiquity*. 36:123-30.

Allan, J. C. 1968. The accumulations of ancient slag in the south-west of the Iberian Peninsula. *Bull. HMG*. 2:47-50.

Allan, J. C. 1970. *Considerations of the antiquity of mining in the Iberian Peninsula*. Occasional Paper No 27. London: Royal Anthropological Institute.

Anstee, J. W., and Biek, L. 1961. A study in pattern welding. *Med. Arch*. 5:71-93.

Anteins, A. K. 1968. Structure and manufacturing techniques of pattern welded objects found in the Baltic States. *JISI*. 206:563-71.

Ashton, Thomas Sutcliffe. 1924. *Iron and steel in the Industrial Revolution*. London: Manchester University Press.

Ashton, Thomas Sutcliffe. 1948. *The Industrial Revolution, 1760-1830*. London: Oxford University Press.

Atkinson, Norman. 1997. *Sir Joseph Whitworth: "The world's best mechanician"*. UK: Alan Sutton Publishing.

Austin, John, and Ford, Malcolm. 1983. *Steel town: Dronfield and Wilson Cammell 1873-1883*. Sheffield, England: Scarsdale Publications.

Bain, Edgar C. 1939. *Functions of the alloying elements in steel*. Cleveland, OH: American Society for Metals.

- http://www.msm.cam.ac.uk/phase-trans/2004/Bain.Alloying/ecbain.html

Baines, Edward. 1835. *History of the cotton manufacture in Great Britain*. London: Fisher, Fisher & Jackson.

Barber, Martyn. 2003. *Bronze & the Bronze Age metalwork and society in Britain c. 2500 - 800 BC*. Stroud, UK: Tempus Publ. Ltd.

Barraclough, K. C. 1973. An eighteenth century steelmaking enterprise: The company of cutlers in Hallamshire, 1759-1772. *Bull. Historical Metallurgy Group*. 6:26-8.

Barraclough, K. C. 1973. The origins of the British steel industry. *Metallurgy Materials Technology*. 623-9.

Barraclough, K. C. 1976. *Sheffield steel*. Derbyshire, UK: Moorland Publishing Co., Ltd. for Sheffield City Museums.

Barraclough, K. C. 1977. Benjamin Huntsman, 1704-1776. *Journal of Historical Metallography Society*. 11:25-9.

Barraclough, K. C. 1979. Early steelmaking in the Sheffield area. *Trans Hunter Archaeological Society*. 10:335-43.

Barraclough, K.C. 1984a. *Steelmaking before Bessemer: Blister steel, the birth of an industry*. Vol 1. London: The Metals Society.

Barraclough, K. C. 1984b. *Steelmaking before Bessemer: Crucible steel, the growth of technology*. Vol 2. London: The Metals Society.

Barraclough, K. C. 1990. Swedish iron and Sheffield steel. *Transactions of the Newcomen Society*. 61:79-80.

Barraclough, K. C. 1991. Steel in the Industrial Revolution. In *The Industrial Revolution in metals*, ed. J. Day and R.F. Tylecote. London: The Institute of Metals.

Barraclough, Kenneth C., and Kerr, J. A. 1976. Steel from 100 years ago. *Historical Metallurgy*. 10.

Beck, Ludwig. 1884-1903. *Die geschichte des eisens in technischer und kulturgeschichtlicher beziehung, von dr. Ludwig Beck*. Braunschweig: F. Vieweg und sohn.

- One of the definitive histories of iron and steel. 5 vols. This important text is not available in an English translation.

Bell, Isaac Lowthian. 1869. *The chemistry of the blast-furnace*. London: Harrison & Sons, Printers.

Bell, Isaac Lowthian. 1886. *The iron trade of the United Kingdom compared with that of the other chief iron-making nations*. London: British Iron Trade Association.

Berdrow, William. 1937. *The Krupps: 150 years Krupp history 1787-1937*. Trans. Fritz Homman. Berlin: Paul Schmidt.

Bergman, Torbern. 1781. *Dissertatio chemica de analysi ferri*. Uppsala, Sweden.

- Extracts from this text are reprinted in C. S. Smith's (1968) *Sources for the history of the science of steel 1532 - 1786*.

Bessemer, Henry, Sir. 1880. *On the manufacture and uses of steel with special reference to its employment for edge tools*. Paper read at Cutlers' Company of London, December 1, 1880.

Bessemer, Henry, Sir. 1905. *Sir Henry Bessemer, F.R.S. an autobiography. With a concluding chapter*. Offices of "Engineering," London.

- Reprinted c. 1989 by the Institute of Metals, Brookfield, VT.

Bexfield, Harold. 1945. *A short history of Sheffield cutlery and the house of Wostenholm*. Sheffield, England: Loxley Bros.

Bick, L. 1978. The archaeological iron and tin cycles. *Archaeophysika*. 10:75-81.

Bining, Arthur Cecil. 1933. *British regulation of the colonial iron industry*. Philadelphia: University of Pennsylvania Press.

Birch, Alan. 1967. *The economic history of the British iron and steel industry 1784-1879*. London: Frank Cass & Co.

Biringuccio, Vannoccio. [1942] 1990. *The pirotechnia of Vannoccio Biringuccio: The classic sixteenth-century treatise on metals and metallurgy.* Trans. and eds Cyril Stanley Smith and Martha Teach Gnudi. Mineola, NY: Dover.

- Extracts from this text are reprinted in C. S. Smith's (1968) *Sources for the history of the science of steel 1532 - 1786.*

Blair, Peter Hunter. 1963. *Roman Britain and early England: 55 B.C. - A.D. 871.* London: Thomas Nelson and Sons Ltd.

Boas, Marie. 1962. *The scientific renaissance 1450 - 1630.* New York: Harper Torchbooks, Harper & Row.

Bober, Harry. 1981. *Jan van Vliet's book of crafts and trades, with a reappraisal of his etchings.* Albany, NY: Early American Industries Association.

- This text contains 18 loose leaf reproductions of van Vliet's etchings of craftsmen at their trades done in the 1630s.

Bohler, R. F. 1908. Tool steel making in Styria. *School of Mines Q.* xxix:329-41.

Boileau, Etienne. 1268. *Le livre des metiers de Paris.* [The Book of Paris Trades.] The original document is in the Bibliotheque Nationale de France.

Bourke, Cormac. 20010). Antiquities from the River Blackwater III, iron axe-heads. *Ulster Journal of Archaeology.* 60.:63-93.

Bracegirdle, Brian. 1973. *The archaeology of the industrial revolution.* London: Heinemann.

Brack, H. G. 1982. *Phenomenology of tools: Philosophical observations on the nature of tool wielding.* West Jonesport, ME: Pennywheel Press.

Brack, H. G. 2008a. *Art of the edge tool.* vol. 7. Hulls Cove, ME: Pennywheel Press.

Brack, H. G. 2008b. *The classic period of American toolmaking1827 - 1930.* vol. 8. Hulls Cove, ME: Pennywheel Press.

Brack, H. G. 2008c. *Registry of Maine Toolmakers.* vol. 10. Hulls Cove, ME: Pennywheel Press.

Brack, H. G. 2008d. *Handbook for ironmongers*. vol. 11. Hulls Cove, ME: Pennywheel Press.

Bradbury, Frederick. 1912. *History of old Sheffield Plate*. Macmillan & Co.

Braidwood, Robert J. 1964. *Prehistoric men*. Glenview, IL: Scott, Foresman and Company.

Breant, J. R. 1823. Description d'un procede al'aide duquel on obtient une espece d'acier fondu, sembable a celui des lames damassees Orientales. *Bull. Societe d'Encouragement pour l'Industrie Nationale*. 22:222-7.

- Reprinted in 1824. *Annales de Mines*. 9:319-28.
- English translation 1824. *Reportory of Arts*. 45:306-14; 1824. *Technical Repository*. 6:49-55; and 1824. *Annals of Philosophy*. 8:267-71.

Brearley, H. 1933. *Steel makers*. London: Longmans Green.

Briggs, Asa. 1979. *Iron Bridge to Crystal Palace: Impact and images of the Industrial Revolution*. London: Thames and Hudson in collaboration with the Ironbridge Gorge Museum Trust.

Brown, Burton T. 1954. *The coming of iron to Greece*. London.

Burke, James. 2007. *Connections*. New York: Simon & Schuster.

Burne, Gordon. 1984. *Iron and steel*. East Sussex, England: Hodder Wayland.

Burnham, Thomas H., and George O. Hoskins. 1943. *Iron and steel in Britain, 1870-1930*. Allen & Unwin.

Caley, E. R. 1964. *Analysis of ancient metals*. Oxford: Pergamon.

Callan, G. B. 1969. 400 years of iron and steel. *Sanderson Kayser Magazine*. 2.

Calvo, F. A. 1963. Metallurgy in Spain. *The Metallurgist*. 2:160-7.

Campbell, R. 1747. *The London tradesman, being a compendious view of all the trades, professions, arts, both liberal and mechanic, now practiced in the cities of London and Westminster. Calculated for the information of parents, and instruction of youth in their choice of business*. London: T. Gardner.

Cantrell, J. A. 1984. *James Nasmyth and the Bridgewater foundry. A study of entrepreneurship in the early engineering industry.* Manchester, UK: Manchester University Press.

Cantrell, John, and Gillian Cookson. 2002. *Henry Maudslay & the pioneers of the machine age.* London: Tempus Publishing Limited.

- An excellent survey of the rise of English engineering, machine design, and construction in the early 19th century.

Carnegie, D. 1913. *Liquid steel: Its manufacture and cost.* London: Longmans Green.

Carpenter, H. C. H. 1928. Alloy steels, their manufacture, properties and uses. *Journal of Royal Society of Arts.* 76:251.

Carr, J. C., and W. Taplin. 1962. *History of the British steel industry.* Oxford: Blackwell.

Ceram, C. W. 1956. *The secret of the Hittites: The discovery of an ancient empire.* Trans. Richard Winston and Clara Winston. New York: Alfred A Knopf.

Charles, James A. (1980). The coming of copper and copper-base alloys and iron: A metallurgical sequence. In: Wertime, Theodore A. and Muhly, James D., Eds. *The coming of the age of iron.* Yale University Press, New Haven, CT. IS.

Charnock, John. 1800-1802. *History of marine architecture.* 3 vols. London: Quarto.

Childe, Vere Gordon. 1925. *The dawn of European civilization.* New York: Knopf.

Childe, Vere Gordon. 1930. *The Bronze Age.* New York: The Macmillan Company.

Childe, Vere Gordon. 1934. *New light on the most ancient East; the oriental prelude to European prehistory.* London: K. Paul, Trench, Trubner & Co., Ltd.

Childe, Vere Gordon. 1944. *The story of tools.* London: Cobbett Publishing Co. Ltd.

Childe, Vere Gordon. 1944. Archaeological ages as technological stages. *Journal of the Royal Anthropological Institute of Great Britian and Ireland.* 74:7-24.

Childe, Vere Gordon. 1947. *Prehistoric communities of the British Isles.* London: Chambers.

Childe, Vere Gordon. 1951. *Man makes himself.* New York: New American Library.

Church, R. A., and E. A. Wrigley, ed. 1994. *The Industrial Revolutions*. 11 volumes. Oxford: The Economic History Society.

- See especially volumes 2 and 3 *The Industrial Revolution in Britain*, volumes 4 and 5 *The Industrial Revolution in Europe*, and volume 6 *The Industrial Revolution in America*.

Clark, Grahame, and Stuart Piggott. 1965. *Prehistoric Societies*. New York: Alfred A. Knopf, Inc.

Clark, J. D. H. 1952. *Prehistoric Europe: The economic basis*. London: Stanford University Press.

Clarke, Helen. 1979. *Iron and man in prehistoric Sweden*. Stockholm, Sweden: Jernkontoret.

Cleere, H. F. 1972. The classification of early iron-smelting furnaces. *Antiquaries Journal*. 52:8-23.

Cleere, Henry, and David Crossley. 1985. *The iron industry of the Weald*. Avon, UK: Leicester University Press.

Clough, R. E. 1985. The iron industry in the Iron Age and Romano-British period. In *Furnaces and smelting technology in antiquity*, ed. P. T. Craddock and M. J. Hughes. British Museum Occasional Paper 48. London: British Museum.

Coghlan, H. H. 1951. *Notes on the prehistoric metallurgy of copper and bronze in the Old World*. Oxford University Press.

Coghlan, H. H. 1977. *Notes on prehistoric and early iron in the Old World*. Pitts-Rivers Museum Occasional Papers on Technology No. 8. Oxford University Press.

Coghlan, H. H., and H. Case. 1957. Early metallurgy of copper in Ireland and Britain. *Proceedings Prehist. Soc.* 23:91-123.

Cooper, Carolyn C. 1981. The production line at the Portsmouth blockmill. *Industrial Archaeology Review*. 6:28-44.

- Portsmouth, England.

Cooper, Carolyn C. 1984. The Portsmouth system of manufacture. *Technology and Culture*. 25:182-225.

Cope, Kenneth L. 1999. *American wrench makers 1830 - 1915*. Mendham, NJ: Astragal Press.

Court, William H. B. 1938. *The rise and fall of the midlands*. London: Oxford University Press.

Craddock, P. T. 1986. Evidence for Bronze Age metallurgy in Britain. *Current Arch.* 9:106-9.

Craddock, P. T. 1995. *Early metal mining and production*. Edinburgh: Edinburgh University Press.

Craddock, P. T. 1998. New light on the production of crucible steel in Asia. *Bulletin of the Metals Museum of Japan Institute of Metals.*

Craddock, P. T., and M. J. Hughes, ed. 1985. *Furnaces and smelting technology in antiquity*. British Museum Occasional Paper 48. London: British Museum.

Cramer, Clayton E. 1995. What caused the Iron Age? *Ancient Near East.*

Cramer, Johann Andreas. 1739. *Elementa artis docimasticae*. London: Leyden.

- Translated into English by Cromwell Mortimer in 1741.
- Extracts from this text are reprinted in C. S. Smith's (1968) *Sources for the history of the science of steel 1532 - 1786.*

Crossley, D. W. 1966. The management of a 16th century ironworks. *Econ. Hist. Rev.* 19:273-88.

Crossley, D. W. 1973. The Bewl Valley Ironworks. *Royal Arch. Inst. Monogr.* London.

Crossley, D. W. 1975. *Sidney ironworks accounts, 1541-1573*. London: Offices of the Royal Historical Society.

Crossley, D. W. 1984. The survival of early blast furnaces: a world survey. *JHMS.* 18:112-31.

Dane, E. Surrey. 1973. *Peter Stubs and the Lancashire hand tool industry*. Altrincham, UK: John Sherratt and Son Lt.

- The best introduction to a very important toolmaker that played a key role in the Industrial Revolution.

- A number of Stubs' tools are in the collection of the Davistown Museum.

Daumas, Maurice, ed. 1969. *A history of technology and invention: Progress through the ages: Volume II: The first stages of mechanization*. New York: Crown Publishers, Inc.

- Originally published as *Histoire Generale des Techniques* in 1964 published by Presses Universitaires de France.

Davies, Oliver. 1935. *Roman mines in Europe*. Oxford: The Clarendon Press.

Day, Joan, and R. F. Tylecote, ed. 1991. *The Industrial Revolution in metals*. Brookfield, VT: The Institute of Metals.

- One of the best texts on the history of metallurgy, indispensable.

Deane, Phillis. 1967. *The first Industrial Revolution*. Cambridge: Cambridge University Press.

della Porta, Giovanni Battista. 1589. Of tempering steel. In *Magiae Naturalis Libri Viginti*, 305-12. Naples, Italy.

- Written in Latin. An anonymous English translation was done in 1658.
- Extracts from this text are reprinted in C. S. Smith's (1968) *Sources for the history of the science of steel 1532 - 1786*.

de Morveau, Louis Bernard Guyton. 1786. Acier. In *Encyclopédie Méthodique. Chymie, Pharmacie et Métallurgie*. vol. 1. Paris, France.

- Extracts from this text are reprinted in C. S. Smith's (1968) *Sources for the history of the science of steel 1532 - 1786*.

Dennis, W. H. 1967. *Foundations of iron and steel metallurgy*. Amsterdam: Elsevier Science.

Derry, Thomas K., and Trevor I. Williams. 1961. *A short history of technology from the earliest times to A.D. 1900*. New York: Dover Publications.

Dickinson, H. W. 1939. *A short history of the steam engine*. Lynchburg, VA: Babcock and Wilcox.

Diderot, Denis. 1959. *A Diderot pictorial encyclopedia of trades and industry: Manufacturing and the technical arts in plates selected from "L'Encyclopédie, ou*

Dictionnaire Raisonné des Sciences, des Arts et des Métiers" of Denis Diderot: In two volumes. New York: Dover Publications Inc.

- One of the most important sources of information on the trades of the early Industrial Revolution.

Diderot, M. and D'Alembert, M. 1751-1777. *L' Encyclopédie ou dictionnaire raisonné des sciences, des arts et des métiers.* 22 vols. Paris: Braissons.

- The Astragal Press has published a reprint of some of this reference set.

Earwood, Caroline. 1993. *Domestic wooden artefacts in Britain and Ireland from Neolithic to Viking times.* Exeter, England: University of Exeter Press.

Emmerling, J. 1978. The technology of Roman swords. *Alt Thuringen.* 15:92-102.

Evans, E. E. 1948. Strange iron objects from county Fermanagh, Ireland. *UJA.* 11:58.

Fairbairn, W. 1859. *On the progress of civil and mechanical engineering.*

Fincham, John. 1851. *A history of naval architecture.* London: Whittaker and Co.

Fisher, Douglas Alan. 1963. *The epic of steel.* NY: Harper & Row.

Friend, J. Newton. 1926. *Iron in antiquity.* London: Charles Griffin and Co.

Forbes, R. J. 1950. *Metallurgy in antiquity: A notebook for archaeologists and technologists.* Leiden, Netherlands: E. J. Brill.

Forbes, R. J. 1955. *Studies in ancient technology.* Leiden, Netherlands: E. J. Brill.

Frere, S. S., ed. 1958. *Problems of the Iron Age in Britain.* Occas. Paper No. 11. London: Inst. Of Archaeology.

Fulford, Michael, David Sim, Alistair Doig, and Jon Painter. 2005. In defense of Rome: A metallographic investigation of Roman ferrous armour from northern Britain. *Journal of Archaeological Science.* 32:241-50.

Gadd, Ian Anders, and Patrick Walles, ed. 2002. *Guilds, society and economy in London, 1450-1800.* London: Center for Metropolitan History.

Gale, N. H., A. Papastamataki, Z. A. Stos-Gale, and K. Leonis. 1985. Copper sources and copper metallurgy in the Aegean Bronze Age. In *Furnaces and smelting technology in antiquity*, ed. P. T. Craddock and M. J. Hughes, British Museum Occasional Paper 48. London: British Museum.

Gale, W. K. V. 1967. *The British iron and steel industry*. England: Newton Abbot.

Gale, W. K. V. 1979. *Iron and steel*. Museum booklet no. 20.05. Shropshire, England: Ironbridge Gorge Museum Trust.

Gardner, J. Starkie. [1892] 1978. *Victoria & Albert Museum: Ironwork: Part I. From the earliest times to the end of the mediaeval period*. Compiled by Marian Campbell. London: Board of Education.

Gardner, J. Starkie. [1896)] 1978. *Victoria & Albert Museum: Ironwork: Part II. Continental ironwork of the Renaissance and later periods*. Compiled by Marian Campbell. London: Board of Education.

Gardner, J. Starkie. [1922] 1978. *Victoria & Albert Museum: Ironwork: Part III. The artistic working of iron in Great Britain from the earlist times*. Compiled by Marian Campbell. London: His Majesty's Stationary Office.

Gilbert, K. R. 1965. *The Portsmouth blockmaking machinery; a pioneering enterprise in mass production*. London: His Majesty's Stationary Office.

Gilbert, K. R. 1971. *Henry Maudslay: Machine builder*. London: His Majesty's Stationary Office.

Gilbert, K. R. 1975. *Early machine tools*. London: His Majesty's Stationary Office.

Gilles, J. W. 1957. The antiquity of the iron industry in the lower Siegerland. *Stahl und Eisen*. 77:1883-4.

Gilles, J. W. 1957. The first twenty five years of research in the Siegerland by digging ancient iron works. *Arch. Eisenh*. 28:179-85.

Gilles, J. W. 1958. New furnace finds in the Siegerland. *Stahl und Eisen*. 78:1200.

Giuseppi, M. S. 1912. Some 14th century accounts of ironworks at Tudeley, Kent. *Arch*. 64:145-64.

Glob, P. V. 1969. *The bog people: Iron-age man preserved*. Trans. Rupert Bruce-Mitford. Ithaca, NY: Cornell University Press.

Goodman, W. L. 1964. *The history of woodworking tools*. New York: David McKay Company, Inc.

- The most important of all publications on woodworking tools and their evolving forms.

Gordon, Robert B. 1987. Sixteenth-century metalworking technology used in the manufacture of two German astrolabes. *Annals of Science*. 44:71-84.

Gordon, Robert B. 1996. *American iron, 1607 - 1900*. Baltimore, MD: Johns Hopkins University Press.

Gordon, R. G., and Reynolds, T. S. 1986. Medieval iron in society. *Technology and Culture*. 27:110-17.

Grabb, J. R. 1975. Shera steel, a forgotten but useful metal. *CEAIA*. 28:8-11.

Grant, Michael. 1969. *The ancient Mediterranean*. New York: A Meridian book, New American Library, NAL Penguin, Inc.

Griffiths, Samuel. [1873] 1967. *Griffiths' guide to the iron trade of Great Britian*. Intro. W. K. V. Gale. Newton Abbott, Devon, England: Davis & Charles.

Grignon, Pierre Clément. 1755. *Mémoires de physique sur l'art de fabriquer le fer*. Paris: Delalin.

- Extracts from this text are reprinted in C. S. Smith's (1968) *Sources for the history of the science of steel 1532 - 1786*.

Guest, Richard. 1823. *A compendious history of the cotton manufacture*. Manchester, England: J. Pratt.

Hadfield, Robert A. 1921. *The work and position of the metallurgical chemist, also references to Sheffield and its place in metallurgy*. London: C. Griffin & Co.

Hamilton, H. 1966. *The Industrial Revolution in Scotland*. London: Frank Cass Publishers.

Harbord, F. W., and J. W. Hall. 1911. *The metallurgy of steel*. London: Charles Griffin.

Harris, J. R. 1978. Attempts to transfer English steel techniques to France in the eighteenth century. *Business and Businessmen.* Liverpool: Marriner, S.

Harrison, Richard J. 1980. *The beaker folk: Copper Age archaeology in western Europe.* London: Thames and Hudson Ltd.

Hayman, Richard. 2005. *Ironmaking: The history and archaeology of the iron industry.* Stroud, UK: Tempus Publishing Ltd.

Hayman, Richard, and Wendy Horton. 2003. *Ironbridge: History & guide.* Stroud, UK: Tempus Publishing Ltd.

Haywood, John, and Barry Cunliffe. 2001. *The historical atlas of the Celtic world.* London; Thames & Hudson.

- A particularly handy overview of early European communities and trading periods.

Healy, J. F. 1978. *Mining and metallurgy in the Greek and Roman world.* London: Thames and Hudson.

Hedges, R. E. M., and C. J. Salter. 1979. Source determination of iron currency bars through the analysis of slag inclusions. *Archaeometry.* 21:161-75.

Henderson, W. O. 1965. *Britain and Industrial Europe 1750 - 1870.* Leicester, UK: Leicester University Press.

Henderson, W. O. 1966. *J. C. Fischer and his diary of industrial England 1814-1851.* New York: A.M. Kelley.

Hermelin, E., E. Tholander, and S. Blomgren. 1979. A prehistoric nickel-alloyed iron axe. *JHMS.* 13:69-94.

Heskel, Dennis, and Carl Clifford Lamberg-Karlovsky. 1980. An alternative sequence for the development of metallurgy: Tepe Yahya, Iran. In *The coming of the age of iron*, ed. Theodore A. Wertime and James D. Muhly. New Haven, CT: Yale University Press.

Higham, Norman. 1963. *A very scientific gentleman: The major achievements of Henry Clifton Sorby.* London: Elsevier.

Hills, Richard. 1970. *Power in the Industrial Revolution.* Manchester, England: Manchester University Press.

Hills, Richard. 1994. *Power from wind: A history of windmill technology.* Cambridge: Cambridge University Press.

Historical Metallurgy Group of the Swedish Ironmasters Association. 1982. *Iron and steel on the European market in the 17th century.* Stockholm, Sweden: Historical Metallurgy Group of the Swedish Ironmasters Association.

Hodgkinson, J. S., and C. F. Tebbutt. 1985. A fieldwork study of the Romano-British iron industry in the weald of southern England. In *Furnaces and smelting technology in antiquity* ed. P. T. Craddock and M. J. Hughes. British Museum Occasional Paper 48. London: British Museum.

Holland, J. 1824. *The picture of Sheffield.* Sheffield, UK.

Holtzapffel, Charles, and John Jacob Holtzapffel. 1846. *Turning and mechanical manipulations.* 5 vols. London: Holtzapffel & Co. Ltd.

- The volume titles are: Volume I: Materials, their choice, preparation and various modes of working them; Volume II: Construction, action, and application of cutting tools; Volume III: Abrasive and other processes not accomplished with cutting tools; Volume IV: Hand or simple turning: Principles and practices; Volume V: The principles & practice of ornamental or complex turning.

Horne, H. 1773. *Essays concerning iron and steel.* London: T. Cadell.

Hulme, E. W. 1943. The pedigree and career of Benjamin Huntsman of Sheffield. *Transactions of the Newcomen Society.* 24:37-48.

Hunter, H. C. 1935. *How England got its merchant marine, 1066-1776.* New York: National Council of American Shipbuilders.

Huntsman, Benjamin, Ltd. 1930. *A brief history of the firm of B. Huntsman Ltd 1742-1930.* Sheffield, England: B. Huntsman, Ltd.

Jackson, Ralph. 1990. *Camerton: The late Iron Age and early Roman metalwork.* London: British Museum Press.

- A detailed description of an important early Iron Age hoard.
- Dr. Jackson presented a copy of this text to the Davistown Museum (1/5/05) while the curator was visiting the British Museum. Many thanks.

James, Simon, and Valery Rigby. 1997. *Britain and the Celtic Iron Age*. London: British Museum Press.

Jars, M. G. 1774-1781. *Voyages metallurgiques*. 3 vols. Lyons, France: L'Academie Royal des Sciences.

Jenkins, Rhys. 1920-21. The rise and fall of the Sussex iron industry. *Transactions of the Newcomen Society*. 1:16-33.

Jenkins, Rhys. 1922-23. Notes on the early history of steel making in England. *Transactions of the Newcomen Society*. 3:16-32.

Jenkins, Rhys. 1936-37. Industries of Herefordshire in bygone times. *Transactions of the Newcomen Society*. 17:178.

Jenkins, Rhys. 1938-39. Ironfounding in England 1490-1603. *Transactions of the Newcomen Society*. 19:35-48.

Jones, G. D. B. 1980. The Roman mines at Rio Tinto. *J. Rom. Stud.* 70:146-65.

Jordan, S. 1889. The iron and steel manufacture in France in 1887, as illustrated by the French exhibits at Paris. *Journal of the Iron and Steel Institute*. 2:10-36.

Jousse, Mathurin. 1627. *La fidelle ouverture de l'art de serrurier*. La Flêche, France.

- English translation by Smith, C. S., and Sisco, A. G. 1961. *Technology and Culture*. 2:131-45.
- Extracts from this text are reprinted in C. S. Smith's (1968) *Sources for the history of the science of steel 1532 - 1786*.

Kilburn, Terence. 1987. *Joseph Whitworth, toolmaker*. Cromford, Derbyshire, UK: Scarthin Books.

Knight, S. A. 1993. The evolution and processes involved in the manufacture of armour plate up to the great war. *Journal of the Ordnance Society*. 5:57-74.

Korb, F., and T. Turner. 1889-90. The manufacture of Styrian open hearth steel. *Journal of the South Staffordshire Institute of Iron and Steel Works Managers*. 6-10.

Kubota, K. 1970. Japan's original steel making and its development under the influence of foreign technique. *Coll. Int. Inst. CNRS*. 538:577-91.

Lance, A. E. 1947. The iron ores of Germany. *JISI*. 156:449-76.

Landes, David. 1969. *The unbound prometheus: Technological change and industrial development in western Europe from 1750 to the present*. Cambridge, UK: Cambridge University Press.

Lang, J. 1995. A metallographic examination of eight Roman daggers from Britain. In *Sites and sights of the Iron Age: Essays on fieldwork and museum research presented to Ian Mathieson Stead*. Barry Raftery, Vincent Megaw and Val Rigby, 119-32. Oxbow Monograph 56. Oxford: Oxbow Books.

Lang, J., P. T. Craddock, and St. J. Simpson. 1998. New evidence for early crucible steel. *Journal of Historical Metallography Society*. 32:7-14.

Lange, E. F. 1913. Bessemer, Goransson, and Mushet: A contribution to technical history. *Manchester Memoirs*. 57:1-44.

Langouet, L. 1984. Alet and cross-channel trade. In *Cross-channel trade between Gaul and Britain in the Pre-Roman Iron Age*, S. Macready and F. H. Thompson, ed, 7-77. Occasional Paper 4. London: Society of Antiquaries of London.

Le Play, F. 1843. Memoire sur la fabrication de l'acier en Yorkshire, et comparison des principaux groupes d'acieres Europeennes. *Annales des Mines*. 4:583-714.

- English translation: Barraclough, K. C. 1973. *Bull. Hist. Met. Group*. 7:14-7.

Lea, Frederick Charles. 1946. *A pioneer of mechanical engineering - Sir Joseph Whitworth*. London: British Council by Longmans, Green.

Livadefs, C. J. 1956. The structural iron of the Parthenon. *JISI*. 182:49-66.

Löwergren, Gunnar. 1948. *Swedish iron and steel, a historical survey*. Trans. Nils G. Sahlin. Stockholm, Sweden: Svenska Handelsbanken.

Maddin, R., J. D. Muhly, and T. S. Wheeler. 1977. How the Iron Age began. *Scientific American*. 131.

Magnusson, G. 1985. Lapphyttan – an example of medieval iron production. In *Medieval iron in society*, ed. Nils Bjorkenstarn, 22-57. Papers presented at the symposium in Norberg, May 6-10, 1985. Sweden.

Maitland, Colonel. 1881. On the metallurgy and manufacture of modern British ordnance. *Journal of the Iron and Steel Institute.* 2:424-51.

Manning, W. H. 1985. *Catalogue of the Romano-British iron tools, fittings and weapons in the British Museum.* London: British Museum Publications Ltd.

Maréchal, J. R., and H. Armand. 1960. *Scientific researches on the metallurgy of the La Tène period and the Roman occupation of Savoy.* Chambery, Annecy: Actes 85th Congr. of Learned Soc.

Martin, Thomas. 1813. *The circle of mechanical arts.* London: Gale Curtis and others.

Mattusch, C. C. 1977. Bronze and iron working; the Forum area. *Hesperia.* 46:380-9.

McGrath, J. N. 1972. A discussion of a reference to steel casting in the early eighteen century. *Journal of the Arms and Armour Society.* 7:53-5.

McHugh, Jeanne. 1980. *Alexander Holley and the makers of steel.* Baltimore, MD: Johns Hopkins Press.

McNeil, Ian. 1968. *Joseph Bramah. A century of invention 1749-1851.* Devon, England: David & Charles PLC.

Merton, Robert K. 1938. Science, technology, and society in seventeenth century England. *Osiris.* 404, 496.

Millard, M. 1911-1912. Old methods of iron making. *Proceedings of the Staffordshire Iron and Steel Institute.* 27:184-96.

Mokyr, Joel. 1993. *The British Industrial Revolution; an economic perspective.* Boulder, CO: Westview Press.

Moore, C. N., and M. Rowlands. 1972. *Bronze Age metalwork in Salisbury Museum.* Salisbury and South Wiltshire Museum Occasional Publication. Salisbury, Wiltshire, UK: Salisbury and South Wiltshire Museum.

Morton, G. R., and J. Wingrove. 1971. The charcoal finery and chafery forge. *Bulletin of the Historical Metallurgy Group.* 5:24-8.

Mott, R. A. 1965. The Sheffield crucible-steel industry and its founder, Benjamin Huntsman. *Journal of the Iron and Steel Institute.* 203:227-37.

Mott, R. A., and P. Singer. 1983. *Henry Cort: The great finer*. London: The Metals Society.

Mouret, Jean-Noël. 1993. *Les outils de nos ancêtres*. France: Hatier.

  • This book is in French, but contains wonderful photographs of "the tools of our ancestors."

Moxon, Joseph. [1703] 1989. *Mechanick exercises or the doctrine of handy-works*. Morristown, NJ: Astragal Press.

Muhly, James D. 1973. *Copper and tin; the distribution of mineral resources and the nature of the metals trade in the Bronze Age*. New Haven, CT: Connecticut Academy of Arts and Sciences.

Muhly, James D. 1973. Tin trade routes of the Bronze Age. *Amer. Scient.* 61:404-13.

Muhly, James D. 1977. New evidence for sources of and trade in Bronze Age tin. In *The search for ancient tin*, A. D. Franklin, et al., ed., 43-8. Washington, DC: Smithsonian Museum.

Muhly, James D. 1980. The Bronze Age setting. In *The coming of the age of iron*, Theodore A. Wertime and James D. Muhly, ed. New Haven, CT: Yale University Press.

Muirhead, James P. 1854. *The mechanical inventions of James Watt*. 3 vols. London: Murray.

Multhauf, Robert R. 1994. Review of *Iron and steel in ancient China*, by Donald B. Wagner. *Technology and Culture: The International Quarterly of the Society for the History of Technology*. 35:605-6.

Mumford, Lewis. 1934. *Technics and civilization*. New York: Harcourt, Brace & Co.

Mushet, D. 1805. Experiments on wootz. *Philosophical Transactions of the Royal Society of London*. 95:163-75.

Musson, A. E. 1978. *The growth of British industry*. Batsford.

Musson, A. E., and E. Robinson. 1969. *Science and technology in the Industrial Revolution*. Manchester, England: Manchester University Press.

Needham, J. 1958. *The development of iron and steel technology in China*. 2 vols. London: Newcomen Society.

Nicholson, John. 1831. *The operative mechanic, and British machinist; being a practical display of the manufactories and mechanical arts of the United Kingdom*. Philadelphia, PA: T. Desilver.

Nicholson, Peter. 1812. *Mechanical exercises*. London.

Nordström, Hans-Åke, and Anita Knape, ed. 1989. *Bronze Age studies: Transactions of the British-Scandinavian Colloquium in Stockholm, May 10-11, 1985*. The Museum of National Antiquities, Stockholm, Studies 6. Stockholm, Sweden: Statens Historiska Museum.

Nuttall, R. H. 1981. The first microscope of Henry Clifton Sorby. *Technology and Culture*. 22.

Oakley, Kenneth P. 1972. *Man the tool-maker*. London: Trustees of the British Museum.

Oddy, W. A., and Judith Swaddling. 1985. Illustrations of metalworking furnaces on Greek vases. In *Furnaces and smelting technology in antiquity*, ed. P. T. Craddock and M. J. Hughes. British Museum Occasional Paper 48. London: British Museum.

O'Hara, J. G., and A. R. Williams. 1979. The technology of a 16th century staff weapon. *Journal of the Arms and Armour Society*. 9:198-200.

Osborn, Fred Marmaduke. 1952. *The story of the Mushets*. London: Thomas Nelson & Son.

Otto, Helmut. 1952. *Handbuch der ältesten vorgeschichtlichen Metallurgie in Mitteleuropa*. Leipzig, Germany: J. A. Barth.

Palmer, Frank R. 1937. *Tool steel simplified: A handbook of modern practice for the man who makes tools*. Reading, PA: The Carpenter Steel Company.

Panseri, C., and M. Leoni. 1966. On the Etruscan technique for making iron arms: Examination of a sword from Montefiascone. *Met. Ital.* 58:381-9.

Parry, G. 1862-1863. On puddled steel. *Proceedings of the South Wales Institute of Engineers*. 3:74-81.

Percy, J. 1864. *The metallurgy of iron and steel*. London: John Murray.

Petree, J. F. 1964. Henry Maudslay - pioneer of precision. *Engineering Heritage*. 1.

Petrie, W. M. Flinders. 1917. *Tools and weapons*. London: British School of Archaeology in Egypt.

Piaskowski, Jerzy. 1982. On the manufacture of high-nickel iron Chalibean steel in antiquity. In *Early Pyrotechnology*, Theodore Wertime. Washington, DC: Smithsonian Institution Press.

- "Antiquity," i.e. c. 1900 BC at the height of the Bronze Age.
- The iron sands of the Black Sea shore of northern Turkey gave rise to a robust nickel steel smelting community.

Piggott, Stuart. 1950. Swords and Scabbards of the British early Iron Age. *Proceedings Prehist. Soc.* 16:1-28.

Piggott, Stuart. 1953. Bronze double-axes in the British Isles. *Proceedings Prehist. Soc.* 19:224-6.

Piggott, Stuart. 1965. *Ancient Europe from the beginnings of agriculture to classical antiquity*. Chicago: Aldine Publishing Company.

Pleiner, Radomir. 1967. *The beginning of the Iron Age in ancient Persia.* Prague: National Technical Museum.

Pleiner, Radomir. 1969. Experimental smelting of steel in early medieval furnaces. *Pamatky archaeologicke.* 458-87.

Pleiner, Radomir. 1969. *Iron working in ancient Greece.* Prague: National Technical Museum.

Pleiner, Radomir. 1979. The technology of three Assyrian artifacts from Khorsabad. *J. Near East. Stud.* 38:87.

Pleiner, Radomir. 1980. Early iron metallurgy in Europe. In *The Coming of the Age of Iron*, ed. Theodore A. Wertime and James D. Muhly. New Haven, CT: Yale University Press.

Pleiner, R., and Judith K. Bjorkman. 1974. The Assyrian Iron Age: The history of iron in the Assyrian civilization. *Proceedings of the American Philosophical Society.* 118:283-313.

Pole, W. 1888. *The life of Sir William Siemens.* Liverpool, UK: John Murray.

Pollard, Sidney. 1905. *Three centuries of Sheffield steel.* Marsh Brothers.

Pulice, Michael. 2006. Machines for making bricks in America, 1800-1850. *The Chronicle.* 59:53-8.

Raftery, Barry, Vincent Megaw, and Val Rigby. 1995. *Sites and sights of the Iron Age: Essays on fieldwork and museum research presented to Ian Mathieson Stead.* Oxbow Monograph 56. Oxford: Oxbow Books.

Raistrick, Arthur. 1938-39. The south Yorkshire iron industry 1698-1756. *Transactions of the Newcomen Society.* 19:51-86.

Raistrick, Arthur. 1953. *Dynasty of iron founders: The Darbys and Coalbrookdale.* New York: Longmans.

Raistrick, Arthur. 1972. *Industrial archaeology.* London: Eyre Methuen.

Randall-MacIver, David. 1925. *Villanovans and early Etruscans.* Oxford: Clarendon Press.

Randall-MacIver, David. 1927. *The Iron Age in Italy.* Oxford: Clarendon Press.

Raymond, Robert. 1986. *Out of the fiery furnace: The impact of metals on the history of mankind.* University Park, PA: Pennsylvania State University Press.

Read, Thomas Thornton. 1932. *Mineral technology; fifteenth century.* Chicago, IL.

de Réaumur, René Antoine Ferchault. 1722. *L'art de convertir le fer forgé en acier.* Paris, France.

- Translated into English as *Réaumur's memoirs on steel and iron* by Sisco, Anneliese G. in 1956.
- Extracts from this text are reprinted in C. S. Smith's (1968) *Sources for the history of the science of steel 1532 - 1786.*

Rees, Jane, and Mark Rees. 1997. *Christopher Gabriel and the tool trade in 18th century London.* Mendham, NJ: Astragal Press.

Rehder, J. E. 1987. The change from charcoal to coke in iron smelting. *Journal of Historical Metallography Society.* 21:37-43.

Rehder, J. E. 1989. Ancient carburization of iron to steel. *Archeomaterials.* 3:27-37.

Renfrew, C. 1971. Carbon 14 and the prehistory of Europe. *Scient. Amer.* 225:63-72.

Renfrew, Colin. 1973. *Before civilization: The radiocarbon revolution and prehistoric Europe*. London: Cambridge University Press.

Rickard, T. A. 1929. Iron in antiquity. *J. Iron Steel Inst.* 120:329.

Rickard, Thomas A. 1932. *Man and metals: A history of mining in relation to the development of civilization.* 2 vols. New York: McGraw-Hill Book Company.

Riley, J. 1885. The rise and progress of the Scotch steel trade. *Journal of the Iron and Steel Institute.* 2:394-403.

Roberts, Kenneth D. 1976. *Tools for the trades and crafts: An eighteenth century pattern book: R. Timmins & Sons, Birmingham.* Fitzwilliam, NH: Ken Roberts Publishing Co.

Roe, J. W. 1916. *English and American tool builders*. New Haven, CT: Yale University Press.

Rostoker, W., and B. Bronson. 1990. *Pre-industrial iron: Its technology and ethnology.* Archeomaterials Monograph No. 1. Philadelphia, PA: Privately published.

Rowlandson, T. S. 1864. *History of the steam hammer*. Manchester, England: Heywood.

Ryan, M., ed. 1978. *The origins of metallurgy in Atlantic Europe*. Dublin: Stationery Office.

Salzman, Louis F. 1913. *English industries of the middle ages*. New York: Houghton Mifflin.

Sandars, N. K. 1978. *The sea peoples: Warriors of the ancient Mediterranean*. London: Thames and Hudson.

Sanderson, C. 1854-55. On the manufacture of steel as carried out in this and other countries. *Journal of the Society of Arts.* 3:450-60.

Sanderson Bros. & Newbould. 1969-71. *400 years of iron and steel.* Sheffield, UK: Sanderson Bros. & Newbould. http://www.sandersonsteel.com/400_years_of_steel.htm

Schoff, Wilfrid H. 1915. The eastern iron trade of the Roman Empire. *Journal of the American Oriental Society.* 35:224-39.

Schubert, John Rudolph Theodore. 1957. *The history of the British iron and steel industry from c. 450 B.C. to A.D. 1775*. London: Routledge & Kegan Paul.

Schubert, H. R. 1951-53. Early refining of pig iron in England. *Transactions of the Newcomen Society*. 28:59-73.

Scientific American. 1976. *Readings from Scientific American: Avenues to antiquity*. San Francisco, CA: W. H. Freeman and Company.

Scrivenor, H. 1854. *History of the iron trade*. London: Longman, Brown, Green & Longmans.

Seebohm, Henry. 1884. On the manufacture of crucible cast steel. *JISI*. 2:372-406.

Sestieri, A. M. Bietti. 1973. The metal industry of continental Italy, 13th to 14th century BC, and its connections with the Aegean. *PPS*. 39:383-424.

Sherby, Oleg D. 1995. Damascus steel and superplasticity – Part I: Background, superplasticity, and genuine Damascus steels. *SAMPE Journal*. 31:10-7.

Siemens, C. W. 1862. On a regenerative gas furnace. *Proceedings of the Institute of Mechanical Eng.* 21-44.

Siemens, C. W. 1868. On the regenerative gas furnace as applied to the manufacture of cast steel. *Journal of the Chemical Society*. 6:279-310.

Siemens, C. W. 1873. On smelting iron and steel. *Journal of the Chemical Society*. 661-78.

Singer, Charles Joseph, ed. 1954-84. *A History of technology*. 8 Vols. Oxford: Oxford University Press.

Sisco, Anneliese G., trans. 1956. *Réaumur's memoirs on iron and steel*. Chicago, IL: University of Chicago Press.

Smiles, Samuel. 1853. *Industrial biography: Iron workers and tool makers*. London: John Murray.

- Reprinted in 1967 with an introduction by L. T. C. Rolt, Newton Abbott: David & Charles.

Smith, Cyril Stanley. 1960. *A history of metallography: The development of ideas on the structure of metals before 1890*. Chicago, IL: University of Chicago Press.

- This is probably the single most important text on the history of ferrous metallurgy as it pertains to Merigovian swords, Damascus steel, and Japanese swords.
- Indispensable.

Smith, Cyril Stanley, ed. 1968. *Sources for the history of the science of steel 1532 - 1786.* Cambridge, MA: The Society for the History of Technology and MIT Press.

- Citations for some of Smith's sources are included this bibliography under: Biringuccio, della Porta, Jousse, de Réaumur, Cramer, Grignon, Bergman, de Morveau, and Vandermonde.
- Another of Smith's indispensable texts, the definitive study of the evolution of steel production in Europe.

Smith, Cyril Stanley. 1981. *A search for structure: Selected essays on science, art, and history.* Cambridge, MA: MIT Press.

Smith, Cyril Stanley, and J. G. Hawthorne. 1974. Mappae Clavicula. A little key to the world of medieval techniques. *Trans. Amer. Phil. Soc.* 64:4.

Smith, Joseph. 1816. *Explanation or key, to the various manufactories of Sheffield, with engravings of each article.* Sheffield, England: J. Smith.

- Reprinted in 1975 by the Early American Industries Association, South Burlington, VT.

Smythe, J. A. 1937-38. Notes on ancient and Roman tin and its alloys with lead. *Trans. Newcomen Soc.* 18:255.

Snodgrass, Anthony M. 1971. *The dark ages of Greece.* Edinburgh: Edinburgh University Press.

Snodgrass, Anthony M. 1980. Iron and early metallurgy in the Mediterranean. In *The coming of the age of iron*, Theodore A. Wertime and James D. Muhly, ed. New Haven, CT: Yale University Press.

Solncev, L. A., R. B. Stepanskaja, L .D. Fomin, and B. A. Sramko. 1969. The earliest objects of cast iron in Eastern Europe. *Sov. Arch.* 1:40-7.

Spencer, T. 1858. On the manufacture of the Uchatius cast steel. *Proceedings of the Institute for Mechanical Eng.* 146-54.

Spring, Laverne W. 1992. *Non-technical chats on iron and steel and their application to modern industry.* Bradley, IL: Lindsay Publications Inc.

Sramko, B. A., L. D. Fomin, and L. A. Solncev. 1977. The initial stages of the working of iron in East Europe. *Sov. Arch.* 1:57-74.

Starley, D. 1999. Determining the technological origins of iron and steel. *Journal of Archeological Science.* 26:1127-33.

Stech-Wheeler, T., J. D. Muhly, K. R. Maxwell-Hyslop, and R. Maddin. 1981. Iron at Taanach and early iron metallurgy in the eastern Mediterranean. *American Journal of Archaeology.* 85:245-68.

Steeds, W. 1969. *A history of machine tools 1700-1910*. London: Clarendon Press.

Steel, David. 1805. *The elements and practice of naval architecture*. London: Quarto.

Stelzer, G. 1959. Ancient German settlement comprising a bloomery hearth and forge at Salzgitter-Lobmachtersen, Siegerland. *Stahl und Eisen.* 79:1201.

Stodart, J., and M. Faraday. 1820. Experiments on the alloys of steel. *Quarterly Journal of Science, Literature and the Arts.* 5:319-80.

Stone, J. F. S. 1960. *Wessex.* In *Ancient Peoples and Places*, ed. Daniel Glyn. New York: Frederick A. Praeger, Inc.

Sutherland, William. 1711. *The shipbuilders assistant, or some essays towards compleating the art of marine architecture*. London.

Swank, James M. 1892. *History of the manufacture of iron in all ages and particularly in the United States from colonial times to 1891*. Philadelphia, PA: The American Iron and Steel Association.

Tebbutt, C. F. 1979. The excavation of three Roman bloomery furnaces at Hartfield, Sussex. *Sussex Archaeological Collections.* 117:47-56.

Tebbutt, C. F., and H. F. Cleere. 1973. A Romano-British bloomery at Pippingford, Hartfield. *Sussex Archaeological Collections.* 111:27-40.

Tholander, Erik. 1971. Evidence of the use of carburized steel and quench hardening in late Bronze Age Cyprus. *Opuscula Atheniensia.* 10:15-22.

Thomas, S. G., and P. C. Gilchrist. 1879. On the elimination of phosphorus. *Journal of the Iron and Steel Institute.* 1:120-34.

Thurston, Robert H. 1878. *History of the growth of the steam engine*. D. Appleton & Co.

Timmins, John Geoffrey, ed. 1984. *Workers in metal since 1784. A history of W. & G. Sissons Ltd.* Sheffield, UK: W. & G. Sissons Ltd.

Timmins, S. 1866. *The resources, products, and industrial history of Birmingham and the Midland hardware district.* London.

Tonsbergh, A. 1982. *Iron and steel on the European market in the 17th century.* Stockholm, Sweden: The Historical Metallurgy Group of the Swedish Ironmasters' Association.

Toynbee, Arnold. 1884. *Lectures on the Industrial Revolution in England*. London: Rivingtons.

- Reprinted in 1956 by Beacon Press, Boston, MA with the title *The Industrial Revolution*.

Truran, W. 1855. *The iron manufacture of Great Britain.* London: E. & F. N. Spon.

Turner, Henry. 1862. *The manufacture of files.* Sheffield, UK: Sheffield Mechanics Institute.

Tweedale, Geoffrey. 1985. Sir Robert Hadfield FRS (1858-1940), and the discovery of manganese steel. *Notes and Records of the Royal Society.* 40:63-73.

Tweedale, Geoffrey. 1986a. *Giants of Sheffield steel: The men who made Sheffield the steel capital of the world.* Sheffield, UK: Sheffield City Libraries.

Tweedale, Geoffrey. 1986b. Metallurgy and technological change: A case study of Sheffield specialty steel and America, 1830-1930. *Technology and Culture.* 27.

Tylecote, Ronald F. 1959. An early medieval iron smelting site in Weardale. *JISI.* 192:26-34.

Tylecote, Ronald F. 1962. *Metallurgy in archaeology; a prehistory of metallurgy in the British Isles*. London: Edward Arnold.

Tylecote, Ronald F. 1970a. Early metallurgy in the Near East. *Metals and Materials.* 285-93.

Tylecote, Ronald F. 1970b. Remains of iron working from the Roman legionary base at Carpow, Perthshire. *Bull. HMG.* 4:3.

Tylecote, Ronald F. 1976. *A history of metallurgy*. London: Metals Society.

- One of the most important publications on metallurgy.
- Indispensable.

Tylecote, Ronald F. 1980. Furnaces, crucibles, and slags. In *The coming of the age of iron*, ed. Theodore A. Wertime and James D. Muhly. New Haven, CT: Yale University Press.

Tylecote, Ronald F. 1982. Metallurgy in Punic and Roman Carthage. In *Mines et fonderies antiques de la Gaule*, ed. C. Domergue, 259-78. Paris: CNRS (National Center for Scientific Research).

Tylecote, Ronald F. 1985. The early history of the blast-furnace in Europe; a case of east-west contact? *Medieval Iron in Society, Norberg Colloq.* 158-73.

Tylecote, Ronald F. 1986. *The prehistory of metallurgy in the British Isles.* London: The Institute of Metals.

Tylecote, Ronald F. 1987. *The early history of metallurgy in Europe*. London: Longmans Green.

Tylecote, Ronald F. 1991. Iron in the Industrial Revolution. In *The Industrial Revolution in Metals*, ed. J. Day and R. F. Tylecote. London: The Institute of Metals,.

Tylecote, Ronald F., and B. J. J. Gilmour. 1986. *The metallography of early ferrous edge tools and edged weapons*. British Archeology Reports No. 155. Oxford: Archeopress.

Ure, Andrew. 1835. *The philosophy of manufactures*. London: Cornmarket Press.

Usher, Abbott Payson. 1905. *An introduction to the industrial history of England.* London: George G. Harrap & Co., Ltd.

Vandermonde, Charles Auguste, Claude Louis Berthollet, and Gaspard Monge. 1788. Mémoire sur le fer consideré dans ses différens états métalliques. In *Mémoires de l'Académie Royale des Sciences*. Paris.

- Extracts from this text are reprinted in C. S. Smith's (1968) *Sources for the history of the science of steel 1532 - 1786.*

Wagner, D. B. 1993. *Iron and steel in ancient China*. Leiden, Netherlands: Brill.

Wainwright, G. A. 1944. Early tin in the Aegean. *Antiq.* 18:57-64.

Waldbaum, Jane C. 1978. *Bronze to iron: The transition from the Bronze Age to the Iron Age in the Eastern Mediterranean.* Studies in Mediterranean Archaeology LIV. Goteborg, Sweden: Paul Åströms Förlag.

Waldbaum, Jane C. 1980. The first archaeological appearance of iron and the transition to the Iron Age. In *The coming of the age of iron*, ed. Theodore A. Wertime and James D. Muhly. New Haven, CT: Yale University Press.

Walton, Mary. 1948. *Sheffield: Its story and its achievements.* Sheffield, UK: Sheffield Telegraph.

Wayman, Michael L. 2000. *The ferrous metallurgy of early clocks and watches: Studies in post medieval steel.* Occasional Paper Number 136. London: British Museum.

- Among the most useful of all references on ferrous metallurgy.

Wertime, Theodore A. 1962. *The coming of the age of steel.* Chicago: University of Chicago Press.

- An important basic reference.

Wertime, Theodore A. 1964. Asian influences on European metallurgy. *Techn. Cult.* 5:391-7.

Wertime, Theodore A. 1973. The beginnings of metallurgy: A new look. *Science.* 182:875-87.

Wertime, Theodore A. 1980a. The pyrotechnologic background. In *The coming of the age of iron*, ed. Theodore A. Wertime and James D. Muhly. New Haven, CT: Yale University Press.

Wertime, Theodore A., and James D. Muhly, ed. 1980b. *The coming of the age of iron.* New Haven, CT: Yale University Press.

Wertime, Theodore A., and Steven F. Wertime, ed. 1982. *Early pyrotechnology: The evolution of the first fire-using industries.* Washington, DC: Smithsonian Institution Press.

- Some very interesting essays.
- A great resource for teachers.

Wheeler, Tamara S., and Robert Maddin. 1980. Metallurgy and ancient man. In *The coming of the age of iron*, ed. Theodore A. Wertime and James D. Muhly. New Haven, CT: Yale University Press.

Whittemore, Laurence F. 1947. *Yankee ingenuity! Then and now*. New York: The Newcomen Society of England American Branch.

Whitworth, J. 1840. *On plane metallic surfaces, and the proper mode of preparing them.*

- This paper is republished in Whitworth (1870).

Whitworth, J. 1870. *Miscellaneous papers on mechanical subjects.* Manchester, UK: G. Faulkner.

Wiener, Martin J. 1981. *English culture and the decline of the industrial spirit 1850-1980.* Cambridge: Cambridge University Press.

Williams, A. R. 1976. Ancient steel from Egypt. *Journal of Archaeological Science.* 3:294-301.

Williams, A. R. 1977a. Methods of manufacture of sword blades in the middle ages. *Gladius.* 13:75-101.

Williams, A. R. 1977b. Roman arms and armour. *J. Arch Sci.* 4:77-87.

Williams, A. R. 1978. Manufacture of armor in 15th century Italy, illustrated by six helmets in the Metropolitan Museum of Art. *Metropolitan Museum Journal.* 13:131-42.

Williams, A. R. 1979. A technical study of some armour of Henry VIII and his contemporaries. *Archeaologia.* 106:157-66.

Williams, A. R., and A. de Reuck. 1995. *The royal armoury at Greenwich, 1515-1649, A history of its technology.* Royal Amouries Monograph 4. London: Royal Armouries.

Wiltzen, T. S., and M. L. Wayman. 1999. Steel files as chronological markers in North American fur trade sites. *Archaeometry.* 41:117-35.

Wing, Anne, and Donald Wing. (date unknown). *The case for Francis Purdew or granfurdeus disputatus.* Self-published.

Wing, Anne, and Donald Wing. 2005. *Early planemakers of London: Recent discoveries in the Tallow Chandlers and the Joiners Companies.* Marion, MA: The Mechanik's Workbench.

Wood, Michael. 1985. *In search of the Trojan War.* New York: Facts on File Publications.

Woodbury, R. S. 1961. *History of the lathe to 1850.* Cambridge, MA: MIT Press.

Wynne, E. J., and R. F. Tylecote. 1958. An experimental investigation into primitive iron smelting technique. *Journal of the Iron and Steel Institute.* 190:339-48.

Zimmer, George Frederic. 1916. The use of meteoric iron by primitive man. *Journal of British Iron and Steel Institute.* 94:343.

Index

144

www.ingramcontent.com/pod-product-compliance
Lightning Source LLC
Chambersburg PA
CBHW051218200326
41519CB00025B/7164